Engineering Science

Engineering Science

H.K.McCorkindale B.Sc., Dip.T.E.

Depute Rector
Banff Academy
Banff

Holmes McDougall · Edinburgh

Holmes McDougall Ltd
Allander House
137-141 Leith Walk
Edinburgh EH6 8NS

ISBN 0 7157 1984-X

Design by Sydney McK. Glen
Illustrations by Hamish Gordon
Picture research by Therese Duriez

Printed in Scotland by Holmes McDougall, Edinburgh

ACKNOWLEDGEMENTS

The author and publishers are grateful to the following for permission to reproduce copyright material:

National Engineering Laboratory: Crown copyright; British Leyland Ltd; Associated Press Ltd; British Railways; Space Frontiers Ltd; Popperfoto; Mowlem Ltd; Timber Research and Development Association; Stothert & Pitt Ltd; The Scotsman Publications Ltd; Scottish Tourist Board; Photograph by British Petroleum; A Shell Photograph; *New Civil Engineer*; United Kingdom Atomic Energy Authority; A. M. Hutcheson and A. Hogg for figure 3.9 from *Scotland and Oil,* Oliver & Boyd; Building Research Establishment; Key Terrain Ltd; Monsanto; John Brown Engineering Ltd; Forth Canoe Club.

Preface

This text-book updates the methods of treatment of problems in engineering science; it encourages graphical methods of solution as well as analytical methods (but which have been included where appropriate).

Throughout the text, emphasis is on a graphical approach — sketches, diagrams and graphs — as well as referring to everyday examples of 'engineering science at work'.

The units used are SI units and particular attention has been paid to the most recent British Standards and also to recommendations regarding Engineering Science Terminology.

Although the context of the text is aimed at the Scottish Certificate of Education examination in Engineering Science at the Ordinary Grade, it will be found to be most useful for students following other engineering courses of similar standing.

H. K. McCorkindale
April 1980

Contents

Chapter 1

Introduction

Everybody has an idea of what an engineer is. What do you think an engineer is?

When you talk about this with your teacher you will find out that the word 'Engineer' means different things to different people. You may think that he or she could be someone who repairs washing machines or someone from the G.P.O. who installs telephones or someone in the garage who services cars. This kind of engineer is a person who repairs machines, fits new machines or looks after machines. But there are many more kinds of engineer and because of this, three broad categories can be used to help define an engineer:

a Skilled Manual Engineer – someone who works mainly with his hands and has served an apprenticeship.

b Technician Engineer – someone who does some manual work and some 'desk' work. He usually needs college qualifications for this.

c Professional Engineer – someone who uses his knowledge of Physics, Mathematics and Engineering Science as well as his experience and creative ability. He is an expert in engineering who solves problems of design. He would normally need a university education as well as good practical experience in order to apply for membership of an Engineering Institution and may become a Chartered Engineer entitled to use the letters C.Eng. after his name.

Different industries need different engineers and again there are several specialist areas which are quite common — for example mining engineering, electronic engineering, aeronautical engineering, production engineering, textile engineering, civil engineering, mechanical engineering, computer engineering and so on. A list like this could go on further. Can you think of a few more?

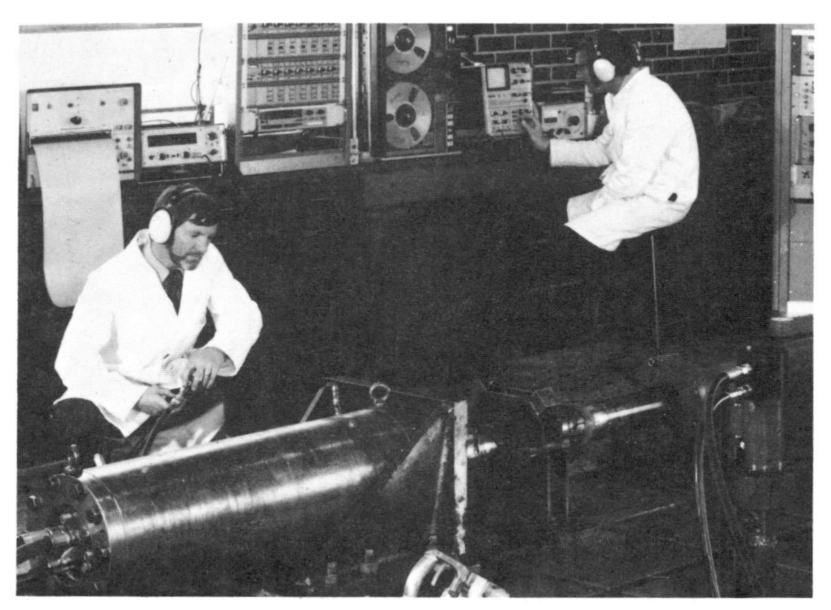

Research and development engineers at work

The subject chosen by one engineer is really his special interest, but all engineers have to use their wits, knowledge and ingenuity to solve particular problems. When Design Engineers are designing something new they have to consider many things – is it strong enough? Can I get enough of the right material? Will it cost too much? Is it the right size? Will it do the job it is supposed to do? Will the production department be able to make it? Is it more complicated than it needs to be? etc.

It doesn't matter whether the thing being designed is a sewing machine needle, an electronic circuit, a bionic limb or a bridge over a river – the same basic questions need to be answered by the engineer.

Look around you at home, at school, in the shops. Almost everything you see has been through the hands of an engineer at sometime. Never before has society been so dependent on technology and engineering and never before have engineers been able to affect society so much. Perhaps one day, some teachers as well as other workers may be replaced by machines.

A car production line fitting cylinder heads

You will have an idea now of how big the field of engineering is. It has something to offer to everyone no matter what their special interest may be, whether it is simply out of interest or as a possible career.

As you work through this course on Engineering Science you will use some of the 'tools' of the engineer – general knowledge, basic mathematics, physics and other sciences, reference books and other publications. You will draw diagrams and make sketches, but perhaps the most important thing is that you will be active in and around the classroom, laboratory and workshop. You will have to observe things carefully, make decisions and note results. The most interesting thing about engineering of any kind is that there is usually more than one correct 'answer' to a problem.

It is very interesting to look back through the years at the engineers of days gone by. So many of them were scientists, mathematicians or artists. (Discuss two or three well known engineers with your teacher including quite recent engineers of note.)

When a builder is building a house, he must always start with the foundations. This is the base on which he builds an attractive interesting structure. The same is true in engineering science. We must start with basics before we move on to even more interesting and exciting applications and the basis of all engineering is measurement. In the next chapter we will see how to measure different things to do with engineering.

Chapter 2

Units and Symbols

Over the centuries man has used many different ways of measuring things. Length or height, for example, have been measured by man at different times by using units such as cubit, ell, hand, pole, vara, etc. More recently in this country we used inch, foot, yard, mile for measuring length. Now, as you know, almost every country in the world is using the metric system – millimetres, metres, kilometres.

There are many things to be measured apart from length or height. For example, how can we measure weight, mass, speed, time, current, heat, etc?

So that we can measure all of these things sensibly a special system of units was devised on the metric system. The official name for this system is the 'Système International d'Unités' (called SI units for short). It is most important nowadays for people to use the same system of units so that everyone knows what 1 metre is and what 1 kilogramme is and so on. These basic units of measurement are very carefully defined and there are people whose job it is to make sure that measuring instruments are accurate.

The five basic quantities which we will be using are length, mass, time, electric current and temperature. The basic units for these are, in order, metre, kilogramme, second, ampere and kelvin.

When we write down different quantities we can use a kind of short-hand. For example, instead of writing the word 'length' we can use 'l'. This is called a **symbol** and in many cases the symbol is just the first letter of the word, sometimes as a small letter and sometimes as a capital letter. We also have symbols for the units, e.g. 'm' for metre.

Basic quantities and their units

Quantity being measured	Symbol for quantity	Unit of measurement	Symbol for unit
length	l	metre	m
mass	m	kilogramme	kg
time	t	second	s
electric current	I	ampere	A
temperature	T	kelvin	K

Therefore, for a bridge 30 metres long we could write $l = 30$ m. For an electric current of 5 amperes we could write $I = 5$ A and so on. The quantities and units shown above are the basic units and we should always try to use these basic units wherever possible. But when we are dealing with very large numbers or very small fractions, the basic unit may not be the best one to use. For example, we measure the distance between towns in kilometres, not metres. 'Kilo' in front of 'metres' means 'thousand', therefore 1 kilometre = 1000 metres.

Again we can use a symbol 'k' for 'kilo' and we can write km for kilometre. Kilo is only one of many ways of changing the basic unit. Any word written in front of a basic unit to change the unit is called a prefix and has a particular numerical value. Some of the most common prefixes are listed below along with their symbols and numerical values.

Prefix	Symbol	Multiplication
giga	G	1000 000 000 or 10^9
mega	M	1000 000 or 10^6
kilo	k	1000 or 10^3
milli	m	1/1000 or 10^{-3}
micro	μ	1/1000 000 or 10^{-6}
nano	n	1/1000 000 000 or 10^{-9}
pico	p	1/1000 000 000 000 or 10^{-12}

We choose a prefix so that the number we have to write down is not too long. For example instead of 30 000 m we would write 30 km or 30×10^3 m. Instead of 0·008 m we would write 8 mm or 8×10^{-3} m. Instead of 6/1 000 000 amperes we would write 6 μA or 6×10^{-6} A.

No matter which prefix we use, we must write the complete unit as one word. The symbols for the complete unit (prefix symbol + basic unit symbol) are written close together, e.g. kilometre = km, milliamp = mA, etc.

Also, when writing numbers with their units we must leave a space between the number and its unit, e.g. 10 m not 10m, or 0·9 km not 0·9km, or 2×10^3 kg not 2×10^3kg, etc.

Examples **1** to **6** below show quantities written using only the basic units of measurement. Give two better ways of writing these quantities, one way using a prefix, the other using a power of ten.

1 2500 m
2 0·0706 m
3 0·000 0006 m
4 0·0102 s
5 0·000 024 s
6 0·085 A

So far we have only looked at basic quantities and their units, but there are many more which are a combination of two or more of the basic ones. The following list shows the quantities we will be using, their symbols, the units of measurement and the unit symbols.

You will see from the list that some symbols are used twice or three times for different quantities, e.g. T for temperature and T for torque. Therefore we must be very careful not to confuse symbols and if in doubt, write the full name of the quantity.

List of quantities and units

Quantity	Symbol for quantity	Unit of measurement	Symbol for unit
acceleration (due to gravity)	g	metre per (second squared)	m/s² or m s⁻²
acceleration (in a straight line)	a	metre per (second squared)	m/s² or m s⁻²
angle	α, β, θ	degree	
angle of friction	ϕ	degree	
area	A	square metre	m^2
centroid co-ordinates	$\bar{x}, \bar{y}$	metre	m
charge	Q	coulomb	C
co-efficient of friction	μ	–	–
current	I	ampere	A
density	ρ	kilogramme per cubic metre	kg/m^3 or $kg\,m^{-3}$
diameter	d	metre	m
displacement (in a straight line)	S	metre	m
distance travelled	S	metre	m
efficiency	η	–	–
elastic modulus (Young's)	E	pascal	Pa
electric charge	Q	ampere-hour	Ah
electromotive force	E	volt	V
energy	E, W	joule or kilowatt hour	J kWh
energy (kinetic)	E_k	joule	J
energy (mechanical work)	W	joule	J
energy (potential)	E_p	joule	J

Quantity	Symbol for quantity	Unit of measurement	Symbol for unit
extension or compression	x	metre	m
Force	F	newton	N
friction force	F_f	newton	N
heat (or heat quantity)	Q	joule	J
latent heat	L	joule	J
length	l	metre	m
mass	m	kilogramme or tonne	kg or t
moment	M, T	newton metre	Nm
normal reaction	R_n	newton	N
potential difference	V	volt	V
power	P	watt	W
pressure	p	newton per square metre or pascal or bar	N/m² or $\mathrm{N\,m^{-2}}$ or Pa or bar
radius	r	metre	m
reaction (force)	R	newton	N
resistance (electric)	R	ohm	Ω
specific heat capacity	c	joule per kilogramme kelvin	J/(kg K) or $\mathrm{J\,(kg\,K)^{-1}}$
speed	v, u	metre per second	m/s or $\mathrm{m\,s^{-1}}$
strain	ϵ	–	–
stress	σ	pascal, or newton per square metre, or newton per square milli-metre	Pa, N/m² or $\mathrm{N\,m^{-2}}$, N/mm² or $\mathrm{N\,mm^{-2}}$
temperature (absolute)	T	kelvin	K
temperature (common)	t	degree Celsius	°C
temperature interval	ΔT	kelvin or degree Celsius	K or °C
time taken	t	second minute hour day	s min h d
torque	T, M	newton metre	Nm
velocity	v, u	metre per second	m/s or $\mathrm{m\,s^{-1}}$
voltage	V	volt	V
volume	V	cubic metre or litre	m³ or l
weight	W, F_g	newton	N
work	W, E	joule	J

Number work

Throughout this book there are worked examples and extra exercises for students to try which involve arithmetic. When trying examples, we should work to only three significant figures; this will be quite accurate enough for our purposes.

For example, we should write 0·333 as the decimal number for $\frac{1}{3}$. Similarly an answer which came to 1·218 km should be written as 1·22 km. For practice, write the following quantities correct to three significant figures using the correct units with prefixes where necessary. (Number 1 is done for you.)

1 24 060 m = 24·1 km or $24{\cdot}1 \times 10^3$ m
2 12 345 m
3 0·0006283 m
4 0·010608 m
5 102·625 A
6 0·020078 A
7 0·0004567 s
8 840·82 s
9 20·0099 A
10 1·205 m

You will now be able to write down very large and very small quantities in a neat and standard form which will be understood by most engineers. Also you will be able to read and understand measurements in the practical work you will be doing. Now we can start on one of the most important topics in engineering science – Dynamics.

Cyclists in the Tour of Italy cycling race

Chapter 3

Engineering Dynamics

Motion is a part of everyday life. We walk to school, run for a bus, go cycling, travel by car, bus, train or plane. Motion can be fast or slow and it can be made faster or slower.

In this chapter we are going to look at how things move and how motion can be changed, that is to say, we are going to study **Dynamics**.

3.1 Definitions

Before we can start our study we must learn some basic definitions so that everyone understands what we mean when we use them.

A **body** is the name which may be used for any object which is being studied. It may be a car, a cyclist, a sledge or a ball.

Distance travelled is the total length of the path actually followed by a moving body no matter how it twists or turns. This quantity is given the symbol S and is measured in metres (m) or kilometres (km).

Displacement is the name given to the length of the shortest straight line path drawn from the starting point of the motion to the finishing point. You can think of this as the 'distance as the crow flies' from start to finish. This quantity is also given the symbol S and is measured in metres (m) or kilometres (km). The precise direction is also very important. For example displacement could be given as '20 km due North' or '6 m vertically down' and so on.

Difference between displacement and distance travelled

Sometimes the size of the displacement is the same as the distance travelled, but this only happens when the actual path is a straight line and the motion is from one end to the other. More often, distance travelled is not the same as displacement, but greater than displacement.

Road maps are a good example of this difference as shown in Diagram 3.1. By measuring the straight line distance from town A to town B we can find the displacement, but by road from A to B the distance travelled will be greater, because it is not a straight line.

It is quite possible for a body to move and yet, at the end of its motion, have no final displacement. This happens when a body finishes its motion at its starting point. For example a car making a trip from town A to town B and back to town A again. At the end of the trip the speedometer shows that the car has been on a journey, but the car is back where it started and so there is no final displacement.

Further examples of this could be an aircraft making a round trip, a pendulum making one complete oscillation or a walk round the house.

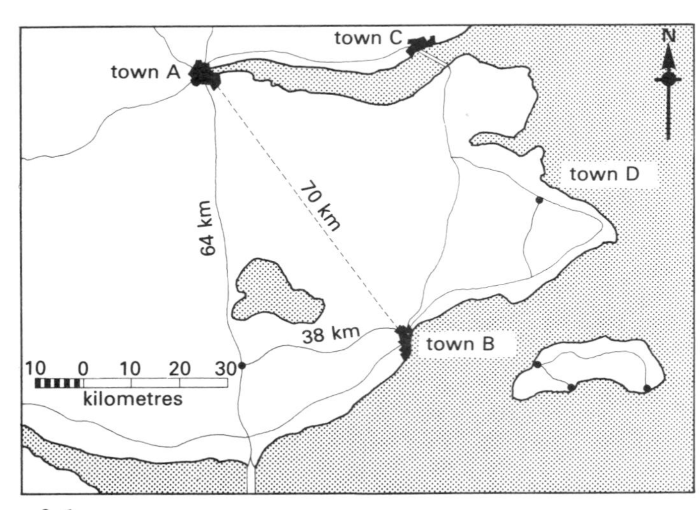

Diagram 3.1

3.2 Distance/time graphs

So that motion can be studied we must be able to record the motion. One way of doing this is to draw a graph of the distance travelled during the time taken. This is called a distance/time graph (S/t graph). It is important to label each axis of the graph with the name of the quantity and its unit of measurement. For example 'Distance in m' and 'Time in s'.

Example 1

A cyclist noted that he travelled a distance of 5 metres every second while free wheeling down a hill. Draw the distance/time graph for the first 4 seconds of his freewheeling.

Solution

The distance/time graph is shown in Diagram 3.2 and this clearly shows that for equal intervals of time (say 1 second intervals) there are equal increments in distance travelled (5 metres).

The kind of motion described in the above example is called **uniform motion**, that is motion which is steady, neither speeding up nor slowing down. For more complicated movement we should use a graph of

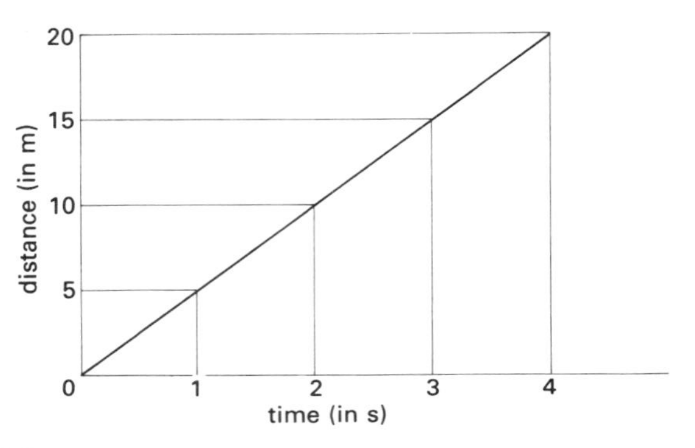

Diagram 3.2

displacement against time, and we must say which direction we will take as being the positive direction. The other direction will therefore be negative. You will have come across this sort of thing already in mathematics when drawing graphs. The x axis and the y axis both have positive and negative directions, and the point where they cross is called the origin.

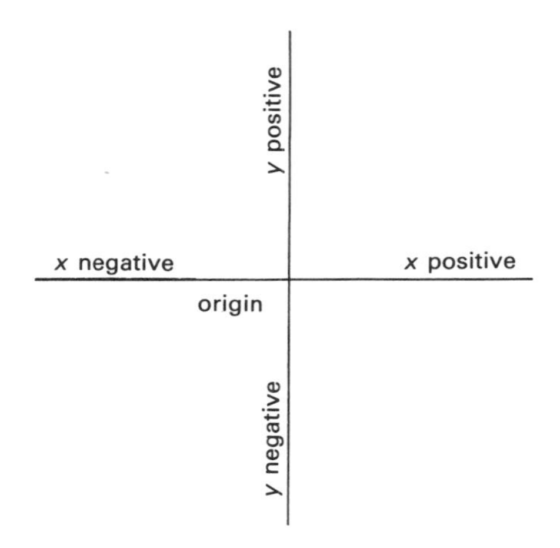

Diagram 3.3

We will use a similar arrangement where the x axis will be time (t) and the y axis will be displacement (S). It will be most unusual to have 'negative time', but whether we have positive or negative displacement really depends on which direction we take as positive to begin with. Therefore the axes for our displacement/time graph (S/t graph) will be drawn as shown in Diagram 3.4.

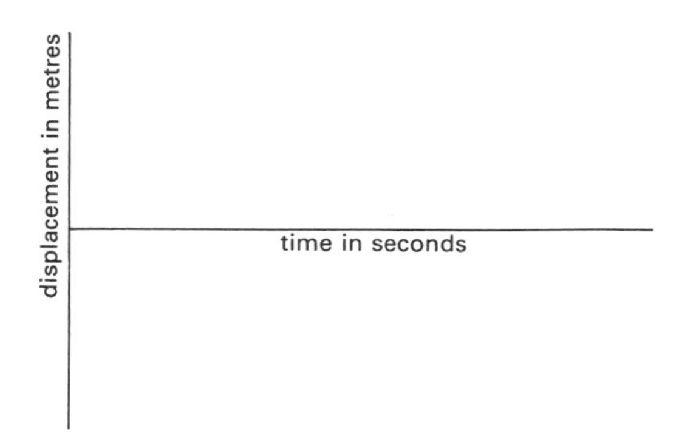

Diagram 3.4

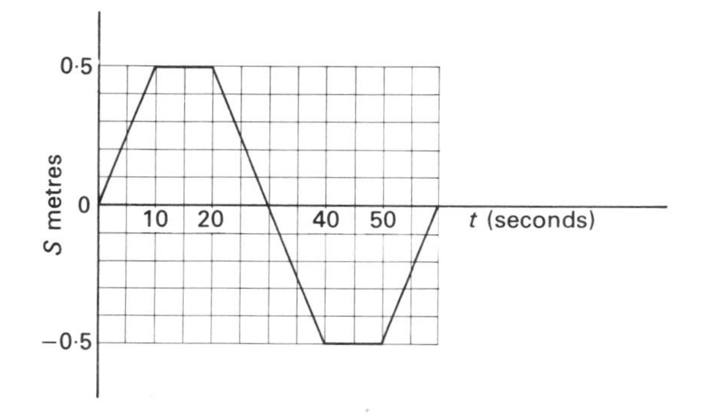

Diagram 3.5

Example 2

A horizontal slide valve is used to close off either a right hand pipe or a left hand pipe. The valve starts from half-way between the pipes and displacement to the right of the origin is called a positive displacement. Diagram 3.5 shows the displacement of the valve over a 60 second period. From the graph, find how far the slide valve is from the origin after:

i 10 seconds
ii 20 seconds
iii 30 seconds
iv 40 seconds
v 50 seconds
vi 60 seconds

Solution

The height of the graph gives the displacement from the origin – upwards from the t axis is positive displacement, i.e. to the right and downwards from the t axis is negative displacement, i.e. to the left of the origin. Therefore:

i after 10 seconds the slide valve is 0·5 m to the right
ii after 20 seconds the slide valve is still 0·5 m to the right
iii after 30 seconds the slide valve is at the origin
iv after 40 seconds the slide valve is 0·5 m to the left
v after 50 seconds the slide valve is still 0·5 m to the left
vi after 60 seconds the slide valve is at the origin

3.3 Speed and velocity

You will already have an idea of what speed is. Perhaps you would say that it is 'how fast' something is moving and you would have the right idea.

In dynamics we would define speed as follows. **Speed** (symbol u or v) is the rate of change of position of a body and is measured in metres per second. In other words it is the distance a body travels in one second:

$$\text{speed} = \frac{\text{distance}}{\text{time}}$$

The units of speed, metres per second, can be written as either m/s or m s^{-1}. Both forms are perfectly correct and you may use either. From here on, to save writing both forms, only m/s will be used, but remember that you can use the ms^{-1} form if you prefer.

British Rail advanced passenger train

The unit of 'metres per second' may not be the best for measuring the speed of a car or a rocket, and so it would be perfectly correct to use kilometres per hour (km/h), as already mentioned in Chapter 2.

However, we must be able to change from metres per second to kilometres per hour and vice versa.

Since 1 km = 1000 m
and 1 h = 3600 s
then $1 \text{ km/h} = \frac{1000}{3600} \text{ m/s}$

(It is best to remember the conversion factor in this form.)

Example 3

Convert the following speeds in km/h to speeds in m/s: **i** 36 km/h; **ii** 72 km/h; **iii** 18 km/h; **iv** 216 km/h.

Solution

i $36 \text{ km/h} = \frac{1000}{3600} \times 36 = 10 \text{ m/s}$

ii $72 \text{ km/h} = \frac{1000}{3600} \times 72 = 20 \text{ m/s}$

iii $18 \text{ km/h} = \frac{1000}{3600} \times 18 = 5 \text{ m/s}$

iv $216 \text{ km/h} = \frac{1000}{3600} \times 216 = 60 \text{ m/s}$

Average speed

Not all motion is steady or uniform, in fact very little is. In a car journey the speed of a car will vary depending on the road conditions, e.g. traffic lights, slow moving traffic, straight clear roads, double bends and so on. At the end of the journey the speedometer will show the total distance travelled and the time taken for the journey can be noted from a clock or stop watch. Some drivers work out their **average speed** for a journey using these two pieces of information.

Average speed (symbol V_{av}) is the value of constant speed which would cause a body to travel the same distance in the same time.

$$\text{Average speed} = \frac{\text{total distance travelled}}{\text{total time taken}} \text{ m/s}$$

in symbols $V_{av} = \frac{S}{t}$

Example 4

A car travels from town A to town B a distance of 192 kilometres in a time of 3 hours. Calculate its average speed.

Solution

$$\text{Average speed} = \frac{\text{total distance travelled}}{\text{total time taken}} = \frac{192}{3} = 64 \text{ km/h}$$

average speed = 64 km/h

Example 5

An athlete runs 1500 metres in 4 min 10 s. Calculate his average speed.

Solution

$$\text{Average speed} = \frac{\text{total distance travelled}}{\text{total time taken}} = \frac{1500}{250} = 6 \text{ m/s}$$

average speed = 6 m/s

Velocity (symbol u or v) is defined as the rate of change of position of a body **in a particular direction**.

In this case the direction is that of the shortest straight line from the start of the motion to the end of the motion, i.e. the direction of the displacement.

$$\text{Velocity} = \frac{\text{displacement}}{\text{time}}$$

The units of velocity are the same as those for speed (m/s or km/h) but the direction must **always** be stated or indicated. For example 60 km/h due East or 25 m/s vertically down.

Average velocity (symbol V_{av}) applies to velocity in the same way that average speed applies to speed.

$$\text{Average velocity} = \frac{\text{total displacement}}{\text{total time taken}}$$

in symbols $V_{av} = \frac{S}{t}$

3.4 Speed/time graphs (*v/t* graphs)

A most useful way of recording patterns of motion is by plotting a graph of speed against time. This shows how the speed changes over a period of time. A practical example of a speed/time graph is the Tachograph which is to be fitted to lorries, linked to the speedometer.

For our work, we will restrict our study to bodies which are at rest, or in a state of uniform motion, or changing speed uniformly or a combination of these.

Examples of speed/time graphs are shown in Diagrams 3.6-3.10.

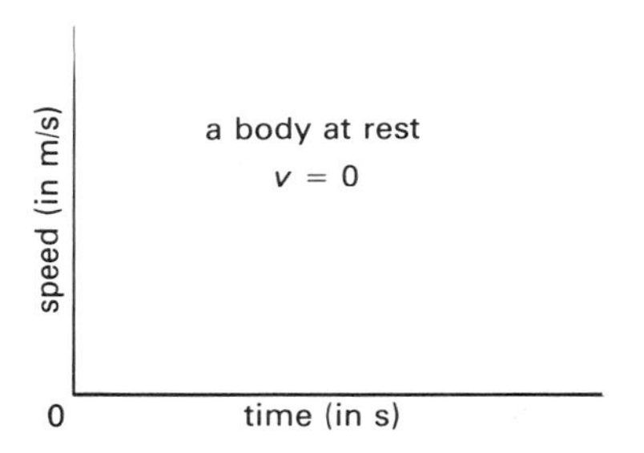

Diagram 3.6

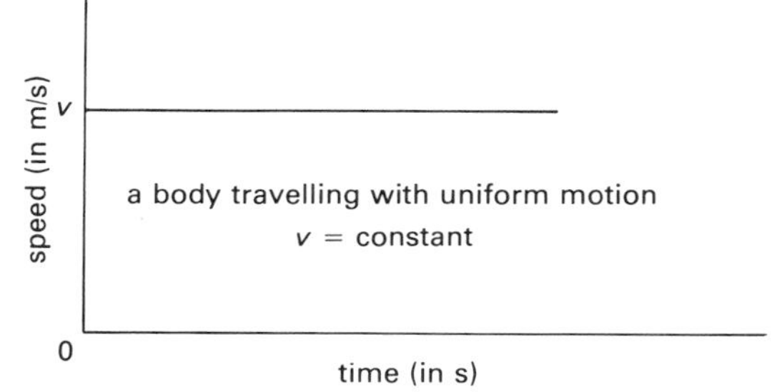

Diagram 3.7

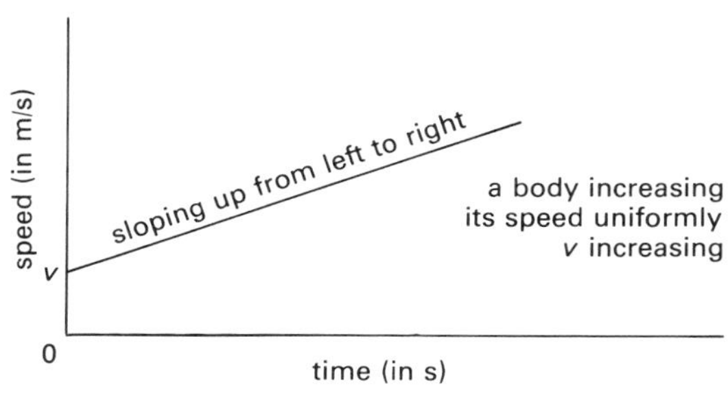

Diagram 3.8

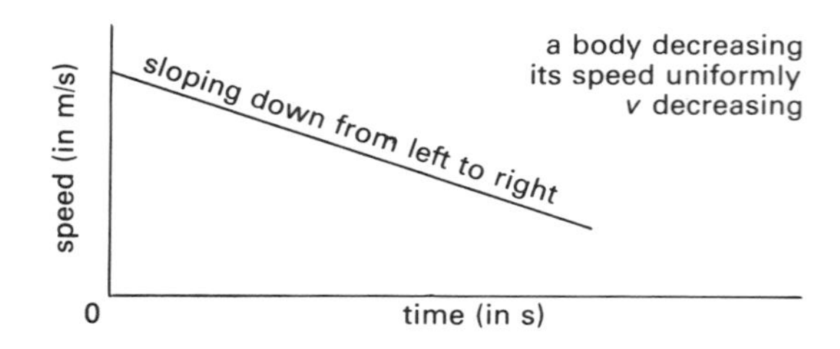

Diagram 3.9

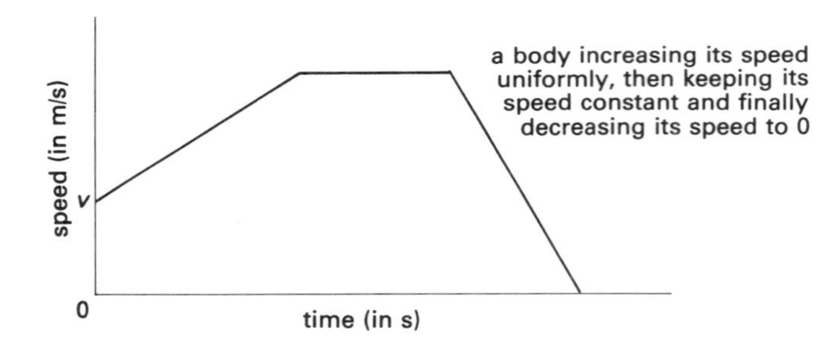

Diagram 3.10

Example 6

Draw the speed/time graph for a cyclist who starts from rest, increases his speed uniformly to 5 metres per second in a time of 10 seconds, maintains this speed for a further 20 seconds and comes to a stop uniformly in another 15 seconds.

Solution

This is in three separate stages:

i increasing speed uniformly from 0 to 5 m/s in 10 s
ii constant speed of 5 m/s for 20 s
iii reducing speed uniformly from 5 m/s to 0 in 15 s

Total time taken = 10 + 20 + 15 = 45 s

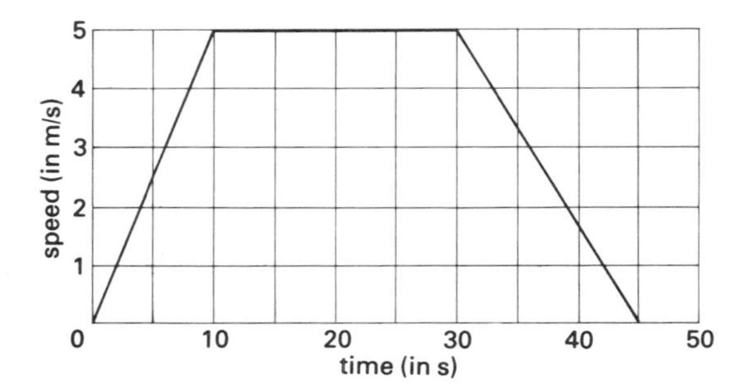

Diagram 3.11

Relationship between distance, speed and time

We have already seen that,

$$\text{average speed} = \frac{\text{total distance travelled}}{\text{total time taken}}$$

This expression can be re-written in another way:

total distance travelled = average speed × total time taken
in symbols $S = V_{av} \times t$

Therefore, if we know the average speed and the total time, then we can work out the total distance travelled. This is quite straightforward for a body travelling with uniform speed because the average speed **is** the uniform speed.

Consider a cyclist travelling with a uniform speed of 5 metres per second for 20 seconds. The total distance travelled would be found as follows:

$$\begin{aligned}\text{total distance travelled} &= \text{average speed} \times \text{total time taken} \\ &= 5 \times 20 \\ &= 100\ \text{m}\end{aligned}$$

To get this answer, we multiplied speed and time together. On the graph of speed/time we can see that by multiplying the speed by the time we get the area under the graph (see Diagram 3.12).

Therefore we have
distance = area of speed/time graph

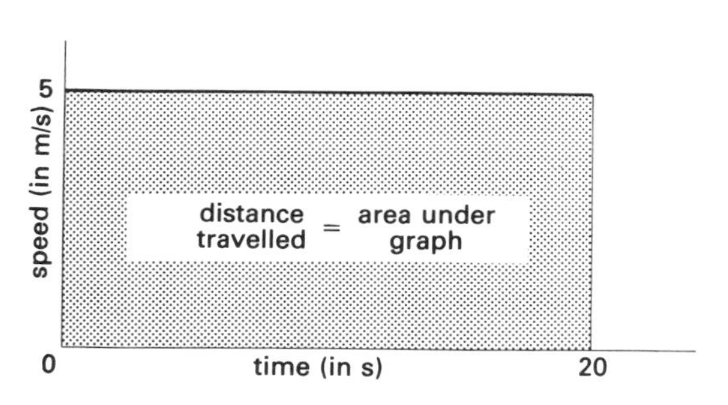

Diagram 3.12

This in fact is always true whether the speed is constant or varying. Diagram 3.13 shows three possible graphs of patterns of motion and in each case the distance travelled can be calculated by simply working out the area under the speed/time graph.

In Example 6 we drew the speed/time graph for a cyclist, as shown in Diagram 3.11. It is now possible not only to draw the speed/time graph, but also to find the distance travelled and therefore the average speed (in Example 6).

To work out the area under the graph it is best to divide it into simple geometrical shapes, in this case two right-angled triangles and a rectangle as shown in Diagram 3.14.

Total distance travelled = total area under graph
= area 1 + area 2 + area 3

(Remember that for any triangle, area of triangle = $\frac{1}{2}$ × base × height.)

total distance travelled = $(\frac{1}{2} \times 10 \times 5) + (5 \times 20) + (\frac{1}{2} \times 15 \times 5)$
= 25 + 100 + 37·5

total distance travelled = 163 m

(Remember that the number part of the answer is best written correct to three significant figures. Therefore the answer appears as 163 m and not 162·5 m.)

We already know that the total time taken was 45 seconds, therefore:

$$\text{average speed} = \frac{\text{total distance travelled}}{\text{total time taken}}$$

$$= \frac{163}{45}$$

average speed = 3·62 m/s (correct to three significant figures)

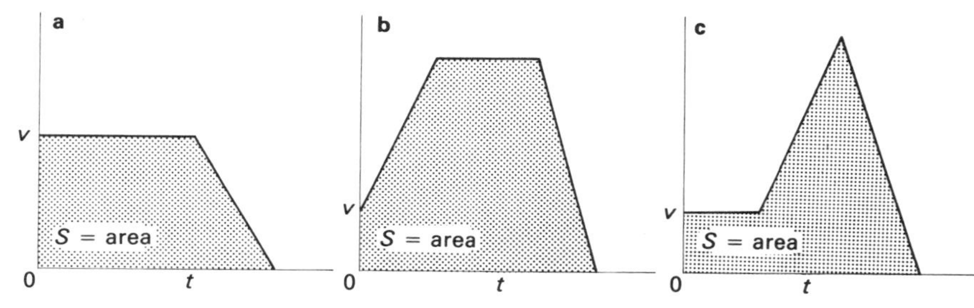

Diagram 3.13

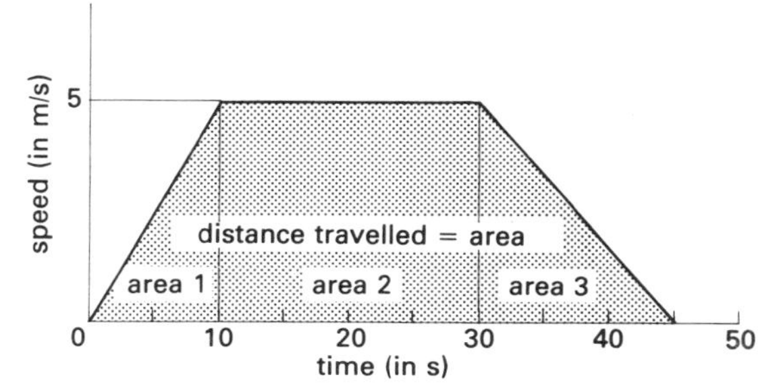

Diagram 3.14

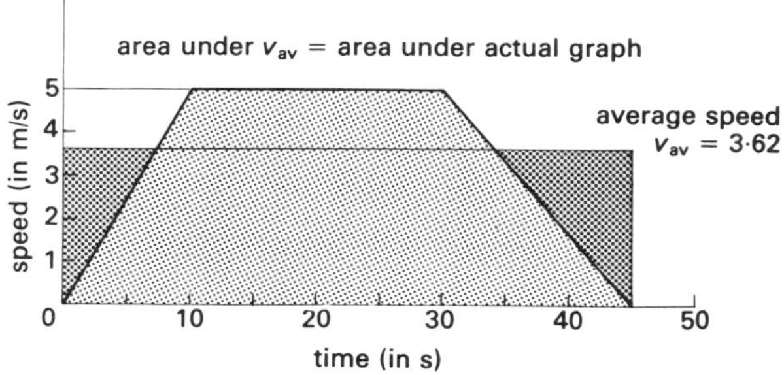

Diagram 3.15

This average speed can be drawn on the original speed/time graph as a horizontal line and the area of the graph below this line will be equal to the area under the original graph (see Diagram 3.15).

Example 7

A car travelling at 18 km/h increases its speed uniformly to 90 km/h in 2 minutes. Draw the speed/time graph and find the total distance travelled (in kilometres) and the average speed (in kilometres per hour) for this period.

Solution

Before we start drawing the graph or doing any calculations, we must be sure that we will not be mixing units. In this example, speed is given in km/h, but time of travel is in minutes. The hours and the minutes cannot be mixed together so we must change the given units into units which we can use together. As a general rule it is best to change all our units into basic units.

Therefore initial speed, 18 km/h $= 18 \times \dfrac{1000}{3600}$

$= 5$ m/s

final speed, 90 km/h $= 90 \times \dfrac{1000}{3600}$

$= 25$ m/s

and time taken, 2 minutes $= 2 \times 60$

$= 120$ s

The graph can now be drawn as in Diagram 3.16. From the graph,

total distance travelled = area under speed/time graph

= area of rectangle + area of triangle

$= (120 \times 5) + (\frac{1}{2} \times 120 \times 20)$

$= 600 + 1200$

$\therefore$ total distance travelled = 1800 m

also, average speed $= \dfrac{\text{total distance travelled}}{\text{total time taken}}$

$= \dfrac{1800}{120}$

$\therefore$ average speed = 15 m/s

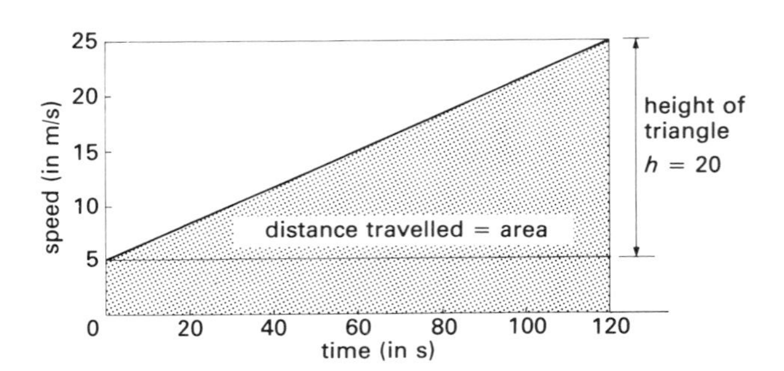

Diagram 3.16

However, the question asks for the answers to be given in km and km/h

$\therefore$ total distance travelled = 1·8 km

and average speed, 15 m/s $= 15 \times \dfrac{3600}{1000}$

$\therefore$ average speed = 54 km/h

3.5 Velocity/time graphs

These are very like speed/time graphs. The difference is that we can find the displacement from the area under the graph. When we drew the displacement/time graph for Example 2 (see Diagram 3.5) we had to take account of positive and negative directions. The same applies to velocities. We must state which particular direction is to be taken as the positive direction and therefore the other direction will be negative.

Displacement = algebraic sum of areas under v/t graph

Example 8

A builder's hoist starts from a landing stage 5 m above the ground and travels vertically upwards in two stages. It increases its upward velocity uniformly from 0 to 4 m/s in 3 seconds and then it decreases its velocity to 0 again in 2 seconds.

It then travels vertically downwards in two stages. It increases its downwards velocity from 0 to 4 m/s in 2 seconds and then it decreases the velocity to 0 again in 4 seconds.

Draw the velocity/time graph for the complete motion, clearly stating which direction is the positive direction for velocity and displacement. Calculate the final displacement and the average velocity.

Solution

Take vertically upwards as the positive direction for both displacement and velocity. Therefore movement upwards will be positive v and positive S, and movement downwards will be negative v and negative S. The graph for the motion of the hoist is as shown in Diagram 3.17. The first triangle on the graph (from 0 to 5 s) is upwards-positive velocity and displacement. The second triangle (5 s to 11 s) is downwards-negative velocity and displacement.

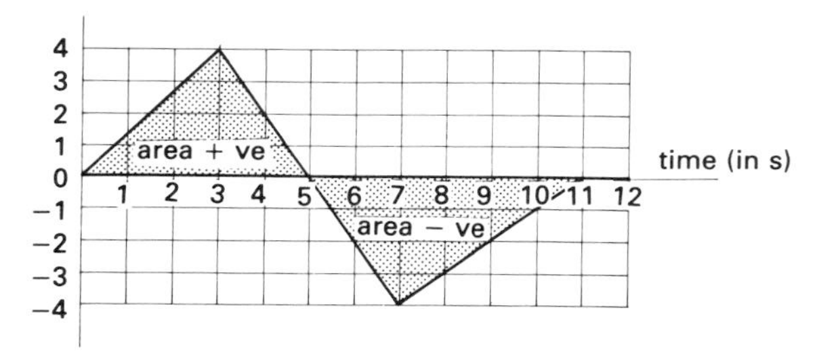

Diagram 3.17

To find the final displacement (as measured from the starting point 5 m above the ground) we must calculate the algebraic sum of the areas under the graph, taking note of 'positive' areas and 'negative' areas.

Final displacement = algebraic sum of areas under v/t graph
= sum of positive areas − sum of negative areas
$= (\frac{1}{2} \times 5 \times 4) - (\frac{1}{2} \times 6 \times 4)$
$= 10 - 12$

final displacement $= -2$ m

N.B. The negative sign means that the final displacement is downwards (below the starting point). This means that after travelling upwards 10 metres and then downwards 12 metres the hoist comes to rest finally 2 m below the point it started from.

Acceleration

In the speed/time and velocity/time graphs, we have used changes in speed and changes in velocity without becoming involved in how big these changes are or how quickly they take place.

A change in speed (or velocity) is more correctly called an acceleration. **Acceleration** is defined as the rate of change of speed (or velocity), i.e. how quickly speed (or velocity) changes from one value to another. Acceleration (symbol a) is calculated by dividing the change in speed by the time taken for the change to take place.

$$\text{Acceleration} = \frac{\text{change in speed}}{\text{time taken}}$$

Therefore the units of acceleration are (metres per second) per second. This unit can be written as $m\,s^{-2}$ or m/s^2, both are equally correct. When reading this unit we would say 'metre per second squared', but remember that it means that the speed is changing by 'so many' metres per second, every second.

For example, a motorcycle accelerating at 2 m/s^2 is actually increasing its speed by 2 m/s, every second. Therefore, if this acceleration carried on for 5 seconds then the speed would change by 5×2 m/s, i.e. 10 m/s change in speed.

Note that acceleration tells us nothing about what speed he started at or what speed he finished at. Acceleration only tells us the rate of **change** of speed.

Mathematically a 'change' in a quantity is found by subtracting the initial value of the quantity from the final value.

Change = final value − initial value
change in speed = final speed − initial speed
or in symbols Δv $= v - u$

therefore acceleration $a = \frac{\Delta v}{t} = \frac{v - u}{t}$

For a body which is slowing down, the initial speed would be larger than the final speed. Therefore the change in speed will be negative and this will make the acceleration negative too.

The negative sign can be ignored and the name acceleration changed to deceleration or retardation.

A typical speed/time graph for a body increasing its speed uniformly is shown in Diagram 3.18. The sloping graph represents the uniformly changing speed, i.e. an acceleration.

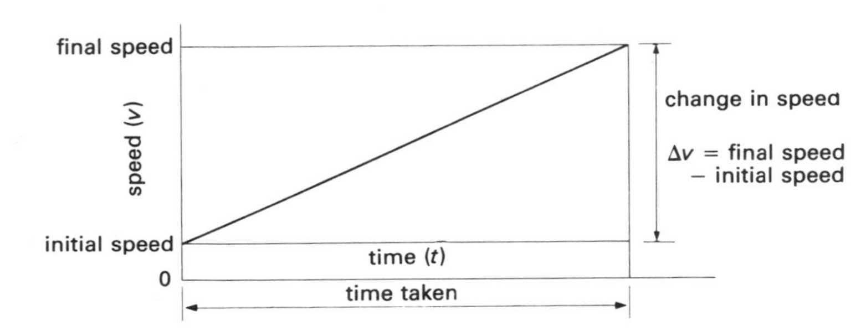

Diagram 3.18

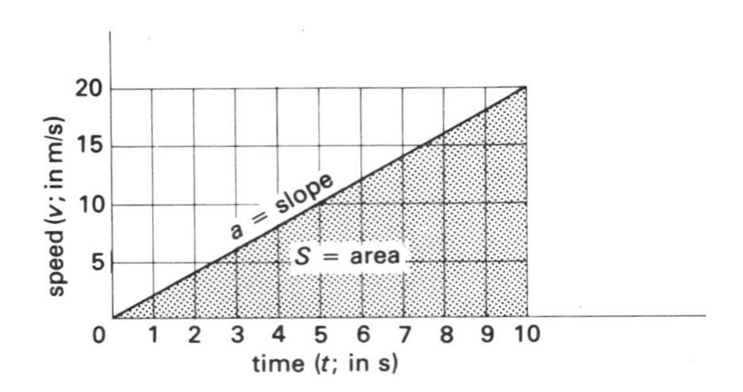

Diagram 3.19

Mathematically the slope (or gradient) of the line is calculated by dividing the vertical change by the horizontal change. On the speed/time graph, the vertical change in the graph is the change in speed and the horizontal change is the time taken.

Therefore slope $= \dfrac{\text{change in speed}}{\text{time}}$

But this ratio is the definition of acceleration, therefore

acceleration = slope of v/t graph

Example 9

A car moves off from traffic lights and increases its speed to 72 km/h in 10 seconds. Draw the v/t graph and find the distance travelled and the average acceleration.

Solution

Initial speed $u = 0$

final speed $v = \dfrac{1000}{3600} \times 72$

$\therefore v = 20$ m/s

time taken $t = 10$ seconds

From the graph in Diagram 3.19

distance = area under v/t graph

$= \frac{1}{2} \times 10 \times 20$

$\therefore$ distance travelled = 100 m

also, acceleration = slope of v/t graph

$= \dfrac{\text{final speed} - \text{initial speed}}{\text{time taken}}$

$= \dfrac{20 - 0}{10}$

$\therefore$ acceleration $= 2\ \text{m/s}^2$

Example 10

After winning a race, at a speed of 144 km/h, the racing car took 1 minute 20 seconds to come to a halt. Draw the v/t graph, find the distance travelled after the end of the race and the retardation.

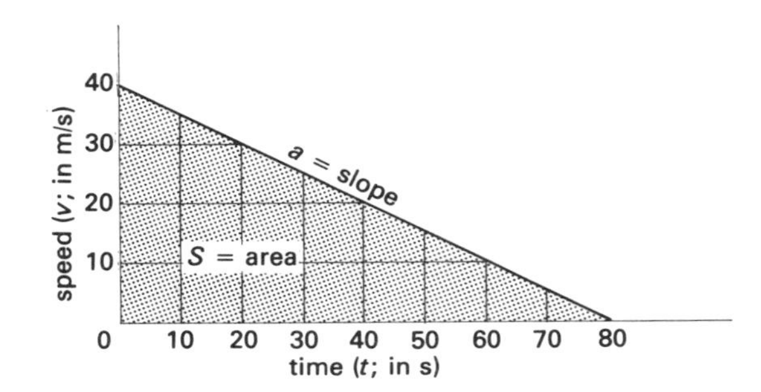

Diagram 3.20

Solution

Initial speed $u = 144$ km/h

$= 144 \times \frac{1000}{3600}$

$= 40$ m/s

final speed $v = 0$

time taken $t = 80$ s

from the graph, distance travelled = area under v/t graph

$= \frac{1}{2} \times 80 \times 40$

$= 1600$ m

∴ distance travelled $= 1{\cdot}6$ km

also, acceleration = slope of v/t graph

$= \frac{\text{change in speed}}{\text{time}}$

$= \frac{\text{final speed} - \text{initial speed}}{\text{time}}$

$= \frac{0 - 40}{80}$

$= -0{\cdot}5 \text{ m/s}^2$

∴ retardation $= 0{\cdot}5 \text{ m/s}^2$

3.7 Vertical motion in the gravitational field

Most people have heard of Sir Isaac Newton who lived from 1642 to 1727. He was a British scientist, astronomer and mathematician who, while drinking tea in his garden one day saw an apple fall. As a result of that very ordinary event, Newton went on to establish the fundamental laws of gravity and motion which in those days were very revolutionary. The reason the apple fell down and not up is because the Earth pulls all free bodies towards its centre.

Before Newton was born, the famous Italian scientist Galileo was experimenting with the effects of gravity. There is a story told that about 1590 Galileo went to the top of the Leaning Tower of Pisa and then dropped two cannon balls of the same size but different weight. He noticed that they fell to the ground at the same time and concluded that motion due to gravity was not affected by weight. The story is no longer believed, but the principle is perfectly correct, as was later proved when scientists were able to construct a tube, and extract nearly all of the air. In this vacuum a coin and a feather were dropped and it was observed that they fell exactly side by side to the bottom of the

The launch of Apollo 11, July 1969

tube. It was also observed that the coin and feather accelerated uniformly as they fell.

No matter how many times the experiment was repeated, the same uniform acceleration was observed. This gravitational acceleration was found to be 9·81 m/s² and it is such an important value that gravitational acceleration is given a particular symbol, g.

$g = 9{\cdot}81$ m/s²

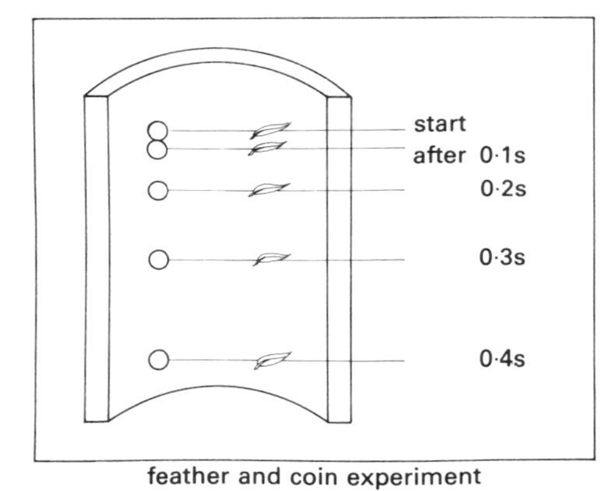

Diagram 3.21

This same experiment was repeated quite recently on the surface of the moon by the astronauts on Apollo 15 in July 1971, but instead of using a coin they used a hammer.

Astronauts during blast-off have to suffer very great accelerations, so great, in fact that they are measured in multiples of normal gravity, g. For example 2 g for 19·6 m/s² or 3 g for 29·4 m/s². Untrained people could become unconscious at about 5 g depending on how long the g-force lasts. Trained astronauts can stand quite large g-forces for quite some time.

Freely falling bodies are quite simply examples of bodies being accelerated uniformly. Therefore the methods of solving problems which we have just discussed should be used, i.e. graphs of speed (or velocity) against time considering the area and slope of the graph. For bodies falling towards the earth, the speed will increase, therefore g is a positive acceleration.

For bodies thrown upwards away from the centre of the earth, their speed will decrease because of gravity and therefore g is a negative acceleration in this case (i.e. a retardation or deceleration).

Example 11

A ball is thrown straight up into the air with a starting speed of 15 metres per second. Calculate how long it will take to reach its maximum height and what that maximum height is.

Solution

The maximum height for the ball is where the ball momentarily stops and then starts falling back to earth again. The graph for this motion starts at 15 m/s at $t = 0$ and the slope of the graph must be the acceleration (in this case $-g$, i.e. downward sloping from left to right).

Let the time taken to reach maximum height be t

i Consider the slope of the graph (see Diagram 3.22).

acceleration = slope of v/t graph

$$= \frac{\text{final speed} - \text{initial speed}}{\text{time}}$$

$$\therefore -9{\cdot}81 = \frac{0 - 15}{t}$$

$$\therefore 9{\cdot}81 = \frac{15}{t}$$

$$\therefore t = \frac{15}{9{\cdot}81}$$

$$= 1{\cdot}53 \text{ s}$$

∴ time taken to reach maximum height = 1·53 s

ii Consider the area under the graph:

maximum height = distance travelled

= area under v/t graph

$= \frac{1}{2} \times 1{\cdot}53 \times 15$

= 11·5 m

∴ maximum height = 11·5 m

Example 12

A marker buoy is dropped from a helicopter hovering at 300 metres above the sea. Find **i** how long the buoy takes to reach the surface of the sea, and **ii** the speed of the buoy as it hits the surface.

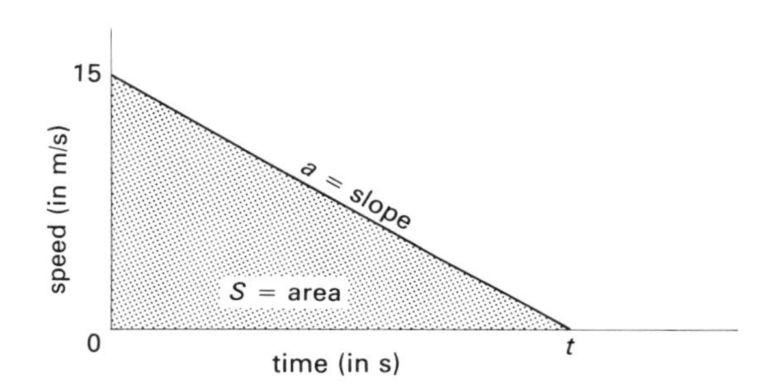

Diagram 3.22

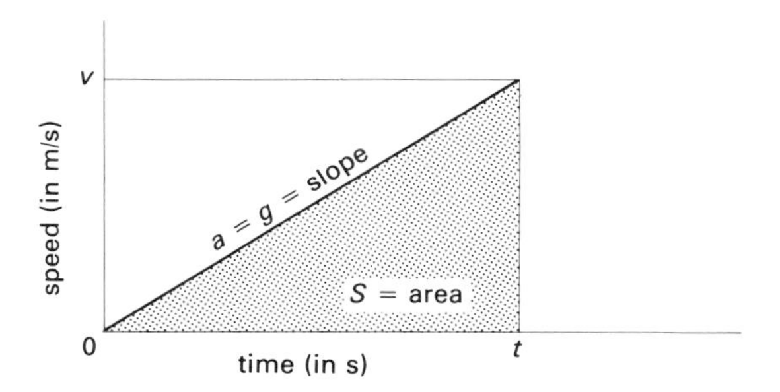

Diagram 3.23

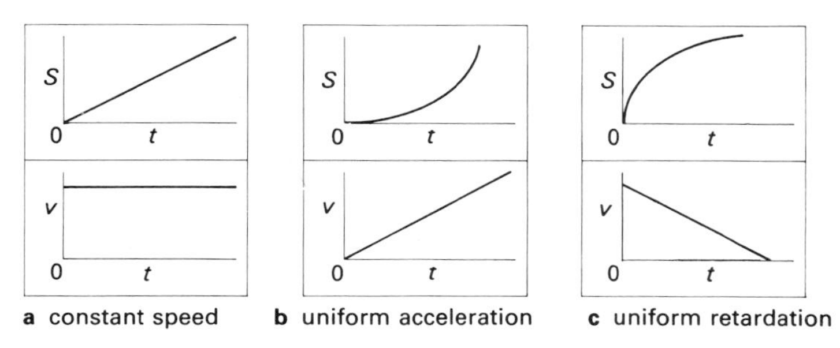

Diagram 3.24

Solution

The v/t graph is as shown in Diagram 3.23

i From the graph, acceleration = slope of v/t graph

$$\therefore g = \frac{v - 0}{t}$$

$$\therefore v = gt \ldots \text{Equation 1}$$

also, distance travelled = area under v/t graph

$$\therefore 300 = \tfrac{1}{2} \times t \times v$$

$$\therefore v = \frac{600}{t} \ldots \text{Equation 2}$$

However, both Equations 1 and 2 are expressions for the same v.

$$\therefore gt = \frac{600}{t}$$

$$\therefore t^2 = \frac{600}{g}$$

$$\therefore t = \sqrt{\frac{600}{g}} = \sqrt{\frac{600}{9{\cdot}81}} = 7{\cdot}82 \text{ s}$$

$\therefore$ time taken to reach maximum height is 7·82 s

ii Using this value of t in equation 1 gives:

$$v = g \times t = 9{\cdot}81 \times 7{\cdot}82 = 76{\cdot}7 \text{ m/s}$$

$\therefore$ speed of buoy at the surface = 76·7 m/s

3.8 Summary of graphs

In all cases:

distance travelled = area under v/t graph
and acceleration = slope of v/t graph

3.9 Analytical methods

An analytical method of solution is one which uses mainly formulae and equations. Therefore we will be using symbols for the quantities:

distance = S
initial speed = u

final speed $= v$
average speed $= V_{av}$
time $= t$
acceleration $= a$

Consider the basic definitions and graphs.

1 From the definition of acceleration,

$$\text{acceleration} = \frac{\text{change in speed}}{\text{time}}$$

$$\therefore a = \frac{v - u}{t}$$

$$\therefore at = v - u$$

$$\therefore v = u + at \text{ . . . Equation 1}$$

2 Consider a typical speed/time graph as shown in Diagram 3.25

distance = area under v/t graph
= area of rectangle + area of triangle

$$\therefore S = ut + \tfrac{1}{2}t\,(v - u)$$

But from Equation 1, $v - u = at$
substituting for $(v - u)$ gives,

$$S = ut + \tfrac{1}{2}at^2 \text{ . . . Equation 2}$$

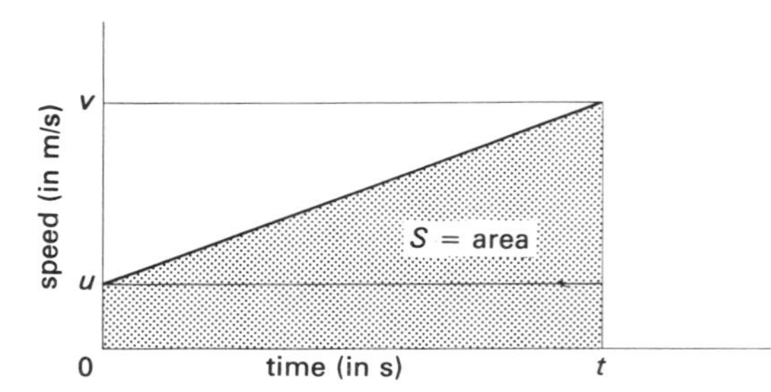

Diagram 3.25

3 Take Equation 1 and square both sides.

$$v^2 = (u + at)^2$$
$$= u^2 + 2uat + a^2t^2$$
$$= u^2 + 2a\,(ut + \tfrac{1}{2}at^2)$$
$$= u^2 + 2aS \text{ (using Equation 2)}$$
$$\therefore v^2 = u^2 + 2aS \text{ . . . Equation 3}$$

4 For a single straight line graph,

$$\text{average speed} = \frac{\text{initial speed + final speed}}{2}$$

$$\therefore V_{av} = \frac{u + v}{2} \text{ . . . Equation 4}$$

5 From the definition of average speed,

$$\text{average speed} = \frac{\text{total distance travelled}}{\text{total time taken}}$$

$$\therefore V_{av} = \frac{S}{t} \text{ . . . Equation 5}$$

Summary of equations of motion

$v = u + at$
$S = ut + \tfrac{1}{2}at^2$
$v^2 = u^2 + 2aS$
$V_{av} = \dfrac{u + v}{2}$
$V_{av} = \dfrac{S}{t}$

Example 13

A cyclist passes the starting line at a speed of 2 metres per second and with an acceleration of 0·25 metres per second squared. Calculate **i** his speed after 8 seconds and **ii** the distance travelled so far.

Solution

From the question we have,

initial speed, $u = 2$ m/s
acceleration, $a = 0{\cdot}25$ m/s²
time, $t = 8$ s

i to find the final speed, v,

using $v = u + at$
then $v = 2 + (0{\cdot}25 \times 8)$
$= 2 + 2$
$= 4$ m/s
$\therefore$ final speed $= 4$ m/s

ii to find the distance travelled, S,

using $S = ut + \frac{1}{2}at^2$

then $S = (2 \times 8) + (\frac{1}{2} \times 0{\cdot}25 \times 8 \times 8)$

$= 16 + 8$

$= 24$ m

$\therefore$ distance travelled $= 24$ m

(N.B. This could also have been done using $V_{av} = \frac{u + v}{2} = \frac{S}{t}$.)

Example 14

An aeroplane standing ready for take-off at one end of a runway 250 metres long needs a speed of 180 kilometres per hour to take off.

Calculate **i** the necessary acceleration and **ii** the time taken for take-off.

Solution

From the question we have,

initial speed, $u = 0$

final speed, $v = 180$ km/h

$= 180 \times \frac{1000}{3600}$

$= 50$ m/s

distance, $S = 250$ m

i to find the acceleration,

using $v^2 = u^2 + 2aS$

then $50^2 = 0 + (2 \times a \times 250)$

$\therefore 2500 = 500a$

$\therefore a = \frac{2500}{500}$

$= 5$ m/s^2

$\therefore$ necessary acceleration $= 5$ m/s^2

ii to find the time taken,

using $v = u + at$ (since we now know $a = 5$ m/s^2)

then $50 = 0 + (5 \times t)$

$\therefore t = 10$ s

$\therefore$ time taken for take-off is 10 s

For freely falling bodies under gravity, the same equations are used but the acceleration, a, will always be $\pm$ 9·81 m/s^2 depending on whether the body is accelerating or decelerating due to gravity.

Example 15

An arrow is shot vertically up into the air with an initial speed of 30 metres per second. Calculate **i** its speed after 2 seconds, **ii** its height after 2 seconds, and **iii** its maximum height.

Solution

From the question, for parts **i** and **ii**

initial speed, $u = 30$ m/s

time, $t = 2$ s

acceleration, $a = -g$

$= -9{\cdot}81$ m/s^2

i to find its speed after 2 s,

$v = u + at$

$= 30 - (9{\cdot}81 \times 2)$

$= 30 - 19{\cdot}6$

$= 10{\cdot}4$ m/s

$\therefore$ speed after 2 s is 10·4 m/s

ii to find height after 2 s,

$S = ut + \frac{1}{2}at^2$

$= (30 \times 2) - (\frac{1}{2} \times 9{\cdot}81 \times 4)$

$= 60 - 19{\cdot}6$

$= 40{\cdot}4$ m

$\therefore$ height after 2 s is 40·4 m

iii to find maximum height, $v = 0$,

$v^2 = u^2 + 2aS$

$\therefore 0 = 30^2 - (2 \times 9{\cdot}81 \times S)$

$\therefore 0 = 900 - 19{\cdot}6\,S$

$\therefore S = \frac{900}{19{\cdot}6}$

$= 45{\cdot}9$ m

$\therefore$ maximum height is 45·9 m

3.10 Force, mass and acceleration

Mass is defined as the amount of matter which makes up a body. Mass is measured in kilogrammes and is given the symbol m. 1000 kg is called 1 tonne (1 t).

The Sunshine Skyway bridge disaster caused by the freighter 'Summit Adventure'

As yet we have not studied force, but generally it is accepted that it is a push or a pull; something which causes things to happen.

We have already mentioned the Earth's pull on all bodies which causes apples to fall with gravitational acceleration. So there is obviously a connection between force and acceleration. Consider two people in a supermarket with their trolleys, one empty the other almost full. They set off round the shop side by side. In terms of acceleration, they both accelerate from rest to a uniform speed (walking speed). The person with the empty trolley will only need to give his trolley a small push because it is empty, but the person with the full trolley not only has to accelerate the trolley but all the goods in the trolley as well, so he needs to give his trolley a very much greater push to keep abreast of his companion.

There obviously must also be a connection between the accelerating force and the amount of matter being accelerated.

It can be shown experimentally that the accelerating force is directly proportional to the mass, i.e. for the same acceleration, twice the mass requires twice the accelerating force, or half the mass requires half the accelerating force, etc. Let the symbol for accelerating force be F_a ('a' is a suffix).

Then $F_a \propto m$ . . . Equation 1

Consider now just one person with a full trolley leaving the supermarket for the car park. As he crosses a roadway he will want to walk more quickly to keep clear of traffic. In terms of acceleration he needs a greater acceleration to reach the same walking speed in a shorter time and this obviously means a bigger push even though the mass of the full trolley is the same as in the supermarket.

Again it can be shown experimentally that for a fixed mass the accelerating force is directly proportional to the acceleration produced, i.e. twice the accelerating force produces twice the acceleration.

$F_a \propto a$ . . . Equation 2

These two expressions 1 and 2 can be combined and in SI units $F_a = ma$. In words, accelerating force = mass × acceleration.

Consider a mass of 1 kg being accelerated at 1 m/s² by an accelerating force F_a.

We know that $F_a = ma$

$\therefore F_a = 1 \times 1$

$\therefore F_a = 1 \text{ kg m/s}^2$

The unit of force (kg m/s²) is given a special name. In SI units it is called the newton (symbol N)

i.e. 1 kilogramme metre per second squared = 1 newton

$1 \text{ kg m/s}^2 = 1 \text{ N}$

We now know how to measure force and therefore we can solve problems about different forces.

Force is defined as that which causes or tends to cause motion. A force of 1 N would cause a body of 1 kg mass to accelerate at 1 m/s² in the direction of the force (there being no resistance). This definition is very useful in determining the Earth's pull on bodies. As has already been shown, freely falling bodies accelerate at 9·81 m/s² towards the centre of the Earth. Consider a body of mass 1 kg freely falling due to the pull of gravity.

Gravitational force causing acceleration = mass × gravitational acceleration
$= mg$
$= 1 \times 9{\cdot}81$
$= 9{\cdot}81 \text{ kg m/s}^2$ or N

∴ the gravitational force on a 1 kg mass = 9·81 N

Because of the importance of gravitational force, it is given a particular symbol F_g (F for force and g due to gravity):

$$F_g = mg$$

The gravitational pull on a body is called the **weight** of the body. 'Weight' is a word which we all use to mean how heavy things are and we tend to think that weight doesn't change. But because of space exploration, we know that there is such a thing as 'weightlessness', which occurs when there is no effective pull of gravity. Also, astronauts on the surface of the moon felt less heavy because the moon's gravity is not as strong as the earth's.

So, although the mass (in kilogrammes) doesn't change, the weight or gravitational force on the mass can change with a change in gravity.

Remember that weight is a force measured in newtons and that

weight = gravitational force
$= F_g$
$= mg$

Examples of weight (on Earth)

The weight of an average eating apple is 1 N.
The weight of a standard bag of sugar is about 10 N.
The weight of a bag of coal is about 500 N.
The weight of an average family car is about 10 000 N or 10 kN.

It is quite interesting that the gravitational pull of the Earth gets less as the distance from the centre of the Earth increases. Because the Earth is not a perfect sphere with uniform thickness of crust, different parts of the Earth, even at sea level have slightly different values of 'g'. Also the higher above sea level the lower is the value of 'g'. These differences are really so small over the surface of the Earth that taking $g = 9{\cdot}81$ m/s² is generally acceptable. On the surface of the moon, the gravitational pull is about one-sixth of the earth's gravitational pull.

A prestressed concrete pile being driven by a diesel pile hammer

Moon's gravity $= \frac{1}{6} \times$ Earth's gravity

$\therefore$ the moon's gravitational acceleration is approximately $1{\cdot}63$ m/s^2

Unless told otherwise, we will assume that all bodies are on the Earth and $g = 9{\cdot}81$ m/s^2.

Example 16

A force of 8 N is applied to a body of 4 kg mass which rests on a perfectly smooth horizontal surface. Find the acceleration.

Solution

$F_a = ma$

$\therefore 8 = 4 \times a$

$\therefore$ a $= 2$ m/s^2

$\therefore$ acceleration is 2 m/s^2

Example 17

An experimental jet car is designed to travel a measured kilometre in 25 seconds. The car has a standing start and accelerates uniformly. The mass of the car is $1{\cdot}5$ tonnes. Draw the v/t graph and find **i** the final speed (in km/h), **ii** the acceleration, and **iii** the accelerating force.

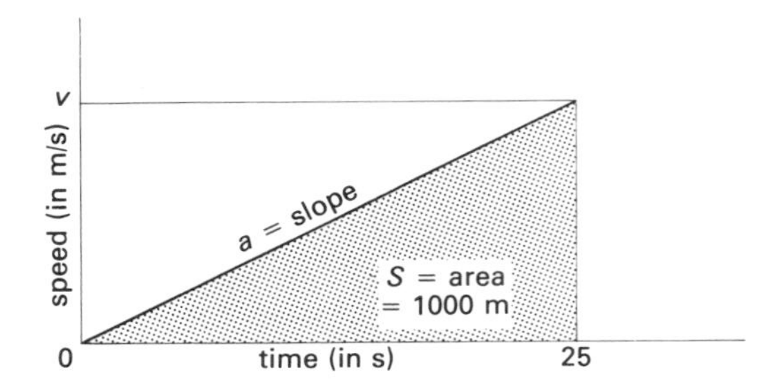

Diagram 3.26

Solution

From the v/t graph shown in Diagram 3.26,

i distance = area under v/t graph

$\therefore 1000 = \frac{1}{2} \times 25 \times v$

$\therefore v = 80$ m/s

changing units from m/s to km/h gives,

$$v = 80 \times \frac{3600}{1000} = 288 \text{ km/h}$$

$\therefore$ final speed is 288 km/h

ii acceleration = slope of v/t graph

$$= \frac{\text{change in speed}}{\text{time}} = \frac{\text{final speed} - \text{initial speed}}{\text{time}} = \frac{80 - 0}{25} = 3{\cdot}2 \text{ m/s}^2$$

$\therefore$ acceleration $= 3{\cdot}2$ m/s^2

iii accelerating force, $F_a = ma$

$$= (1{\cdot}5 \times 1000) \times 3{\cdot}2 = 4800 \text{ N}$$

$\therefore$ accelerating force $= 4{\cdot}8$ kN

Exercises

1 A bus travels between two towns at an average speed of 40 km/h. If the journey takes 45 minutes what is the distance between the towns?

2 A curling stone travels 50 m at an average speed of 2 m/s. How long did its journey take?

3 A cyclist travelling downhill, increases his speed from 9 km/h to 36 km/h in 1 minute. Find **i** his average speed (in m/s), **ii** the distance travelled in that minute (in m), and **iii** his acceleration (in m/s^2). Draw the speed/time graph for this motion.

4 A boy on a sledge starts off at a speed of 8 m/s and comes to a halt after 120 metres. Draw the speed/time graph and find **i** how long his sledge run lasted, and **ii** the deceleration.

5 The graph of speed/time for a vehicle testing its brakes is shown in Diagram 3.27. Its initial speed was 72 km/h and the distance travelled during braking was 50 metres. Complete the graph and find **i** the time taken to come to a halt, and **ii** the deceleration.

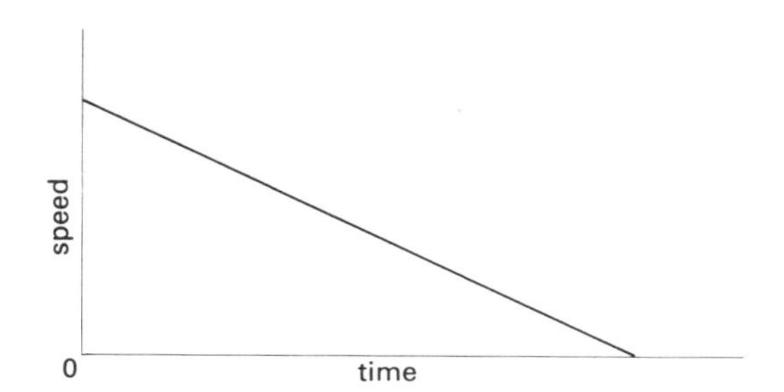

Diagram 3.27

6 A car is known to be able to decelerate at 2 m/s^2. Find how far the car will travel when coming to a stop from a starting speed of **i** 54 km/h, and **ii** 90 km/h.

7 A pebble is dropped down a well which is 60 m deep. How long will it take the pebble to reach the bottom and what will be its maximum speed?

8 A boy throws his football up to a friend standing on a platform, 8 metres above him. What is the lowest speed he must give the football so that it reaches his friend?

9 A search plane flies due North for 2 minutes at 180 km/h then due East for 3 minutes at 200 km/h and finally due South for a further 2 minutes at 180 km/h. Find **i** the distance travelled, **ii** the displacement, and **iii** the average speed.

10 The speed/time graph for the first 7 seconds of a motorcycle journey is shown in Diagram 3.28. From the graph find **i** the distance travelled during the first 5 seconds, **ii** the distance travelled during the next 2 seconds, **iii** the total distance travelled, **iv** the average speed over the 7 second period, and **v** the acceleration during the first 5 seconds.

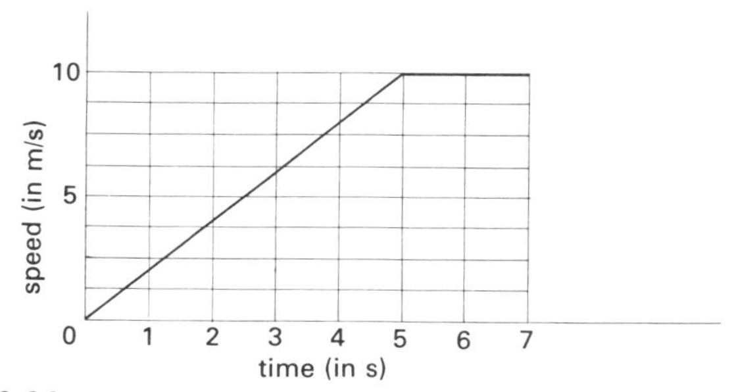

Diagram 3.28

11 A car accelerates at 1·2 m/s^2 from a speed of 6 m/s to a speed of 24 m/s. It then decelerates at 1·8 m/s^2 back to its initial speed of 6 m/s. Draw the speed/time graph and find **i** the time taken for the acceleration, **ii** the time taken for the deceleration, **iii** the total distance travelled, and **iv** the average speed.

12 The speed/time graph for a car changing up through its gears is shown in diagram 3.29. From the graph find **i** the acceleration from 0 to 2 s, **ii** the acceleration from 3 to 7 s, **iii** the acceleration from 8 to 16 s, and **iv** the distance travelled for the 20 seconds of travel shown.

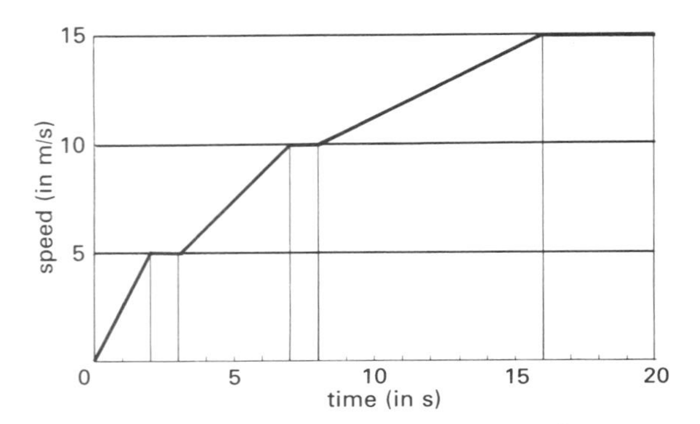

Diagram 3.29

13 A train leaves station A with an acceleration of 0·25 m/s² for 2 minutes. It then travels at this final speed for 10 minutes before decelerating for 3 minutes to bring the train to a stop at station B. Draw the speed/time graph and find **i** the speed at the end of the acceleration (in km/h), **ii** the distance travelled during acceleration, **iii** the distance travelled at constant speed, **iv** the distance travelled during deceleration, **v** the distance between stations A and B, **vi** the average speed (in km/h), and **vii** the deceleration (in m/s²).

14 During the last lap of a car race the leading car, 'A' was 500 m from the finishing line travelling at its maximum speed of 240 km/h. In second place car 'B' was 20 m behind 'A' and was also travelling at 240 km/h, but has a maximum speed of 280 km/h which it just reached as it crossed the finishing line. Who won the race, 'A' or 'B' and by what distance?

15 Find the force necessary to make a mass of 15 kg accelerate at 1·6 m/s² across a perfectly smooth horizontal surface.

16 Determine the weight of a mass of 3 kg.

17 Find the acceleration produced by a force of 12 N applied to a mass of 8 kg.

18 What mass can be accelerated at 2 m/s² by an accelerating force of 60 N?

19 A car of 1·5 tonnes mass is accelerated from rest to a speed of 36 km/h in 4 seconds. Find the necessary accelerating force.

20 A railway engine of mass 25 tonnes travelling at 2 m/s runs into the buffers at a station and is brought to rest in a distance of 0·8 m. Find the average decelerating force of the buffers.

Chapter 4

Statics

In this section of the work we will be looking at bodies which are at rest, even with some forces acting on them. We have seen that a force can cause acceleration, but force can do other things too.

A force can change the direction a body is moving in, for example a footballer heading the ball.

A force can change the shape or size of a body, for example squeezing a ball of wet clay or stretching a spring. Sometimes, however, a force does not make any of these things happen, perhaps because the force isn't big enough.

So we can say that force causes or tends to cause acceleration or change of direction of motion or change of size or change of shape.

4.1 Representing a force

There are two things about a force which are very important. These are:

i the size of the force (called the 'magnitude')
ii which way the force acts (called the 'direction').

So, for any force, we must know its magnitude and its direction. We can draw a line with an arrow to represent a force. The length of the line will show the magnitude of the force. The way the arrow points will show the direction of the force. However, force is measured in newtons, and the length of a line on a page is measured in millimetres. We get over this problem by using a **scale**. We make up a suitable scale so that every millimetre means a certain number of newtons. For example a scale of 1 mm to 5 N means that a force of 500 N would be drawn as a line 100 mm long. For a scale of 1 mm to 10 N, a force of 750 N would be drawn as a line 75 mm long.

When we know the direction of the force as well as the magnitude, we can actually draw the line and arrow in the proper way. The line which we draw is called a **vector** and a diagram which has such lines in it is called a **vector diagram**.

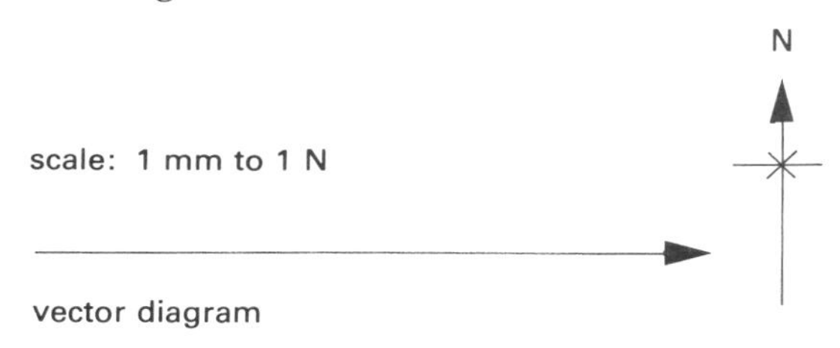

Diagram 4.1

Example 1

A force of 60 N acts due East. Draw the vector for this force on a vector diagram to a scale of 1 mm to 1 N.

Solution

Check this vector by measuring it for yourself.

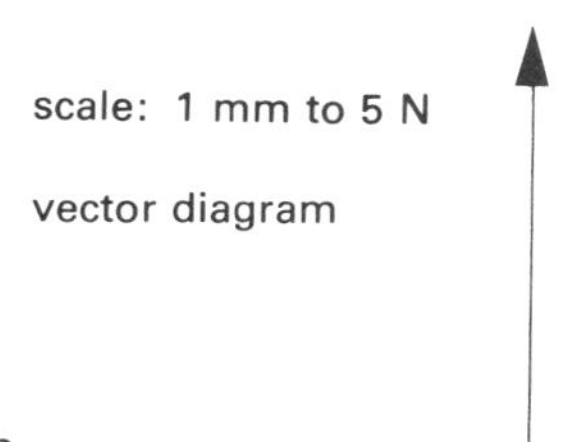

Diagram 4.2

Example 2

A force of 150 N acts vertically up. Draw the vector for this force on a vector diagram to a scale of 1 mm to 5 N.

Solution

Check this vector for yourself.

Example 3

A force of 700 N acts horizontally to the left. Draw the vector for this force on a vector diagram to a scale of 1 mm to 10 N.

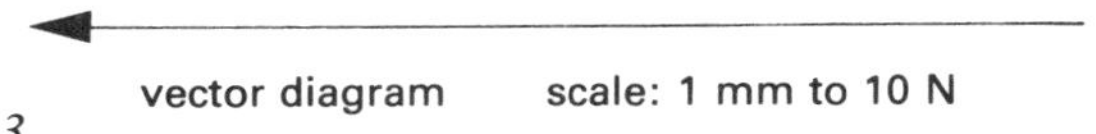

vector diagram scale: 1 mm to 10 N

Diagram 4.3

Solution

Again check this vector.

In all of these examples, the direction of the force was given in words but it is more common for a diagram to be used which shows the force acting on the body. For example look at Diagram 4.4. To describe this situation in words, we could say 'a crate is pulled to the right along a horizontal surface by a force of 100 N which acts at an angle of 30 degrees up from the horizontal to the right'.

Obviously it is easier to understand the sketch of what is happening, so in most cases a sketch will be given to show the forces with their directions. Such a sketch should be called a **space diagram**, because it shows the positions of the body, the forces and the surfaces 'in space'. When drawing space diagrams, the length of the arrows for the forces is not really important because these arrows are not vectors.

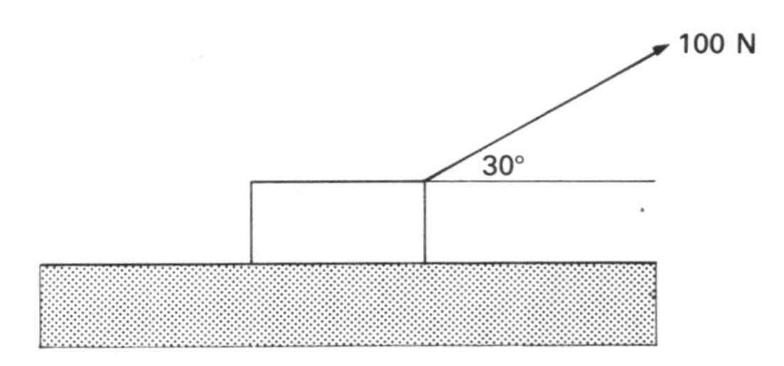

Diagram 4.4

Example 4

A force acts on a body as shown in the space diagram in Diagram 4.5. Draw the force as a vector on a vector diagram to a scale of 1 mm to 2 N.

Solution

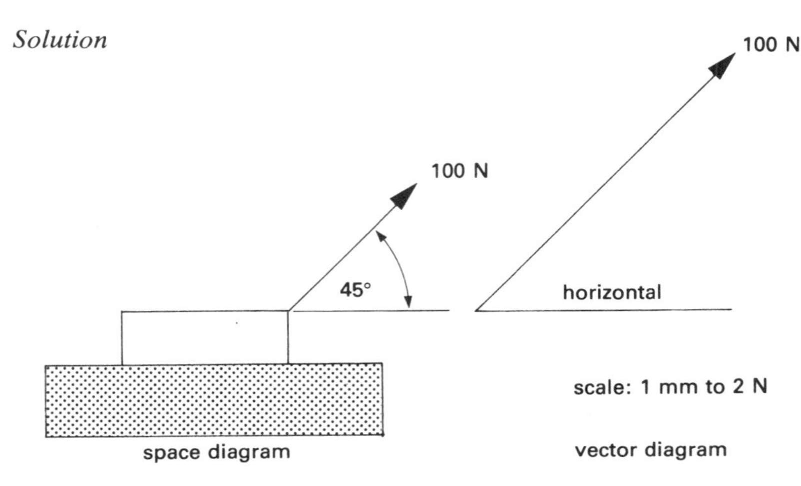

Diagram 4.5

Example 5

A force of 8 N acts on a body as shown in the space diagram in Diagram 4.6. Draw the force as a vector on a vector diagram to a scale of 1 mm to 0·2 N.

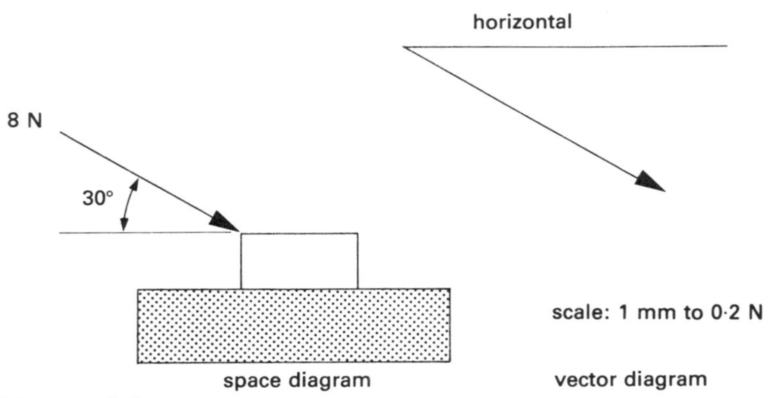

Diagram 4.6

Scales

You will have noticed that the scales used for the vector diagrams are not always the same. The scale is very important and it should be chosen so that the vector diagram is quite large. If we draw very small vector diagrams it is difficult to measure the vectors accurately.

Almost any scale can be used, but it is better to use scales like:

1 mm to 1 N
1 mm to 2 N
1 mm to 5 N
1 mm to 10 N

The scale should always say what 1 millimetre represents. The number part for the force should be:

0·01, 0·1, 1, 10, 100, or
0·02, 0·2, 2, 20, 200, or
0·05, 0·5, 5, 50, 500.

The units of force could be N or kN or MN.

Further examples of scales are:

1 mm to 0·02 N
1 mm to 50 kN
1 mm to 2 MN

When trying to decide what scale to use think of how much space you have on your sheet of paper and make your vector diagram as large as possible.

4.2 Adding forces together

When more than one force acts on a body we need to know what the combined effect will be. In a tug-of-war this is quite easy. The men on one side all pull on the same rope in the same direction, so the total pull for the team is the sum of all the forces – by simple addition.

Example 6

It took two people to push a car along a road. Each person exerted a force of 400 N in the same direction on the back of the car. What was the total force on the back of the car?

Solution

Since both people are pushing in the same direction, the total force is found by simple addition.

$$\text{Therefore the total force} = 400 + 400 = 800\text{ N}$$

When we are adding forces together we are in fact adding vectors. The rule for adding vectors is that each vector in turn must be drawn so that it starts where the last vector finished. This means that in a vector diagram, the vectors are joined 'nose to tail' rather like links in a chain.

In Example 6, the two 400 N forces were in the same direction. Let us assume that the direction was horizontally to the right. Therefore we can draw the vector diagram as shown in Diagram 4.7.

The overall straight line from the start of the first vector to the finish of the second vector is the answer to the vector addition. The length of the line gives the magnitude of the overall force and the direction of the line is the direction of the overall force. Even for forces which are not in the same direction the same rules apply.

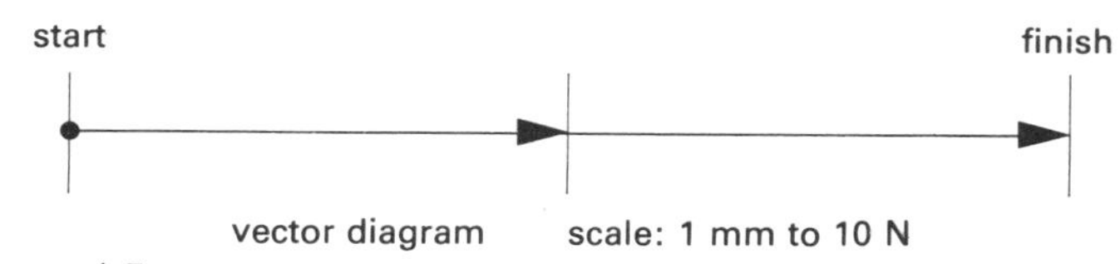

Diagram 4.7

Example 7

Two men slide a crate along a horizontal floor. One man, at the front, pulls on a rope at an angle of 30° up from the horizontal with a force of 260 N. The other man, at the back, pushes down on the crate with a force of 260 N but at an angle of 30° to the horizontal as shown in Diagram 4.8. Find the effective force applied to the crate.

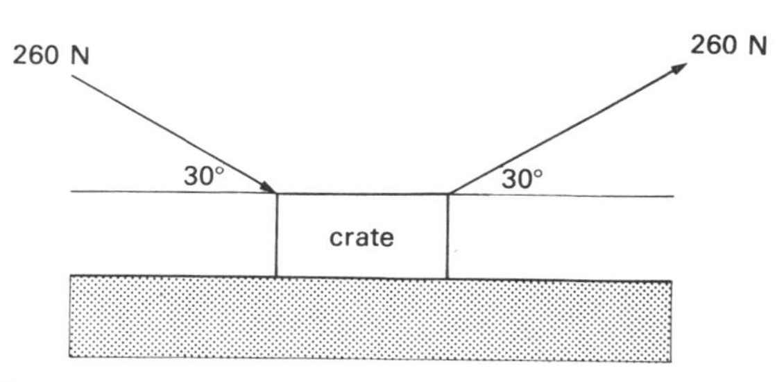

Diagram 4.8

Solution

Because the two forces are not in the same direction we must draw a vector diagram and find the overall straight line from the start to the finish of the vectors.

No matter which force we start with when drawing the vector diagram, we get the same result. In this example, the overall straight line is horizontal from left to right and it is 90 mm long. From the scale of 1 mm to 5 N, the effective force must be 5 × 90 = 450 N, and it acts horizontally to the right. Therefore the effective force = 450 N.

The correct name for this effective force, or overall force is **resultant**.

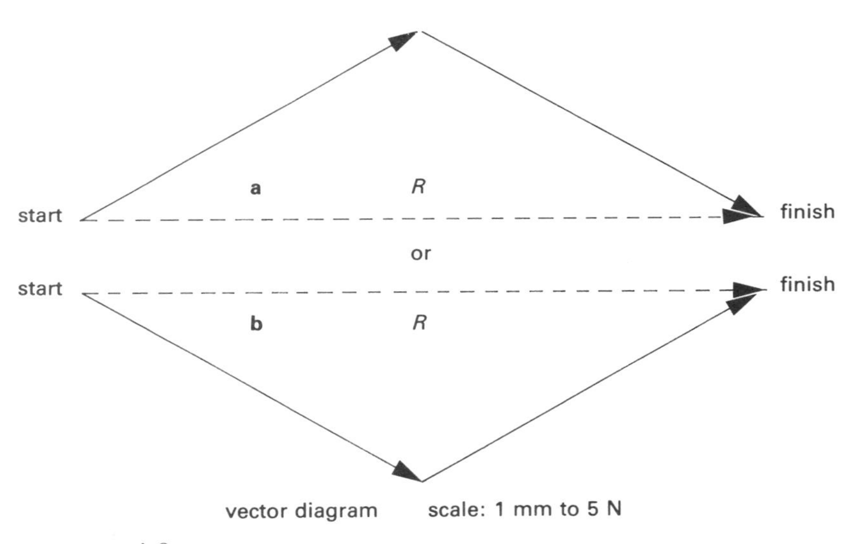

Diagram 4.9

In engineering science we say that the **resultant** is that single force which replaces a system of forces and produces the same effect as the system it replaces. Therefore, on a vector diagram, the single straight line from the start to the finish of the vectors, represents the resultant for that system of forces.

Example 8

Two garage mechanics, fitting a gearbox to a car, found that one had to push vertically up with a force of 185 N while the other had to push horizontally with a force of 400 N towards the front of the car. Find the resultant force on the gearbox.

Solution

Diagram 4.10**a** and **b** show the two ways that the vector diagram can be drawn. Both are correct and give the same answer.

Therefore, resultant = 440 N, 25° ⦫

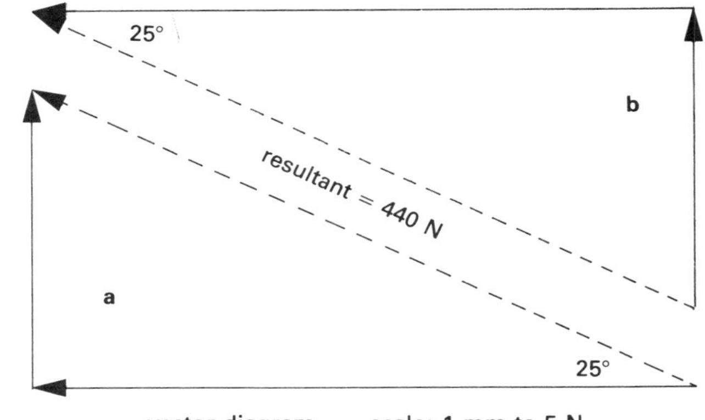

Diagram 4.10

4.3 Resolution of a force into two rectangular components

We have seen how to add two forces together to make a single resultant force. We can do the same in reverse, that is, a single force can be replaced by two other forces provided that these two other forces add up to the single force we started with. These 'other' forces are called **components** and if they are at right angles to each other they are called **rectangular components**. Replacing a force with components is called **resolution** of a force. Therefore the heading for this section means 'replacing a force with two other forces which are at right angles to each other'.

Example 9

Replace the force shown in Diagram 4.11 with two rectangular components.

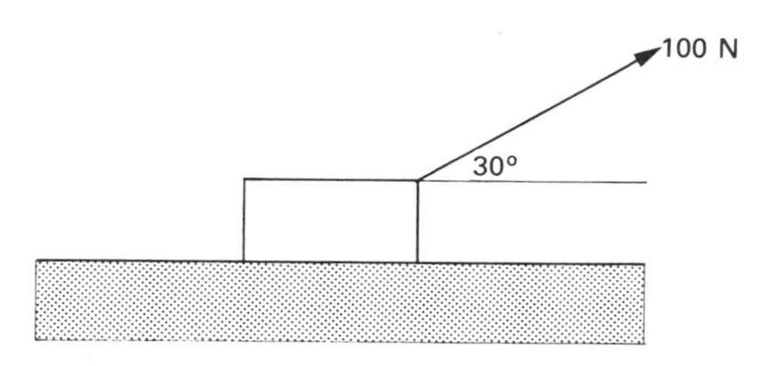

Diagram 4.11

Solution

There are an infinite number of correct answers to this problem because any pair of forces must be correct if they are at right angles to each other and they add up, on the vector diagram, to the given force. Diagram 4.12 shows a few of the correct answers.

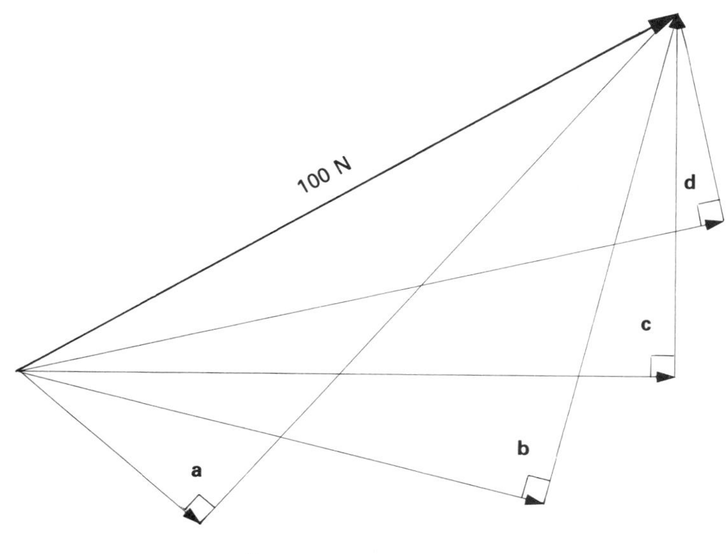

Diagram 4.12

However, the most commonly used rectangular components are those which lie horizontally and vertically. Therefore, for Example 9, the rectangular components most likely to be used would be:

86·5 N horizontally to the right and 50 N vertically up.
(Note: these are measured from the vector diagram as accurately as possible.)

Example 10

Find the horizontal and vertical components of a 100 N force which acts up to the right at 45° to the horizontal.

Solution

By measuring the components to scale in the vector diagram (Diagram 4.13), we find that:

horizontal component = 71 N →
and the vertical component = 71 N ↑

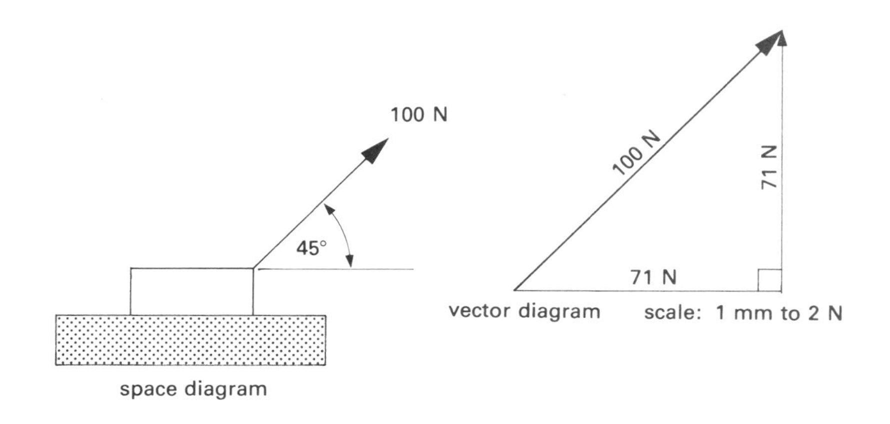

Diagram 4.13

4.4 Equilibrium and equilibrant

'Equilibrium' is the word used in engineering science to mean 'balanced forces'. A body which has several forces on it which all balance each other is said to be in a state of equilibrium, i.e. there is no resultant force making the body accelerate. A body in equilibrium can be at rest or travelling with uniform motion in a straight line.

If there is no resultant force, then on a vector diagram the vectors must finish back at the starting point of the vector diagram, i.e. the vector diagram must 'close'.

For two forces to balance each other, they must be exactly equal and opposite in direction. For example, a perfectly matched tug-of-war contest ending in a draw.

Two forces which are not exactly equal and/or not exactly opposite can not be in equilibrium. Another force would need to be added in order to produce equilibrium. This extra force which would produce equilibrium is called the **equilibrant** and it will be equal in magnitude, but opposite in direction to the resultant force for a system of forces.

Example 11

Diagram 4.14 shows a body under the action of two forces. Find **i** the resultant of these two forces, and **ii** the equilibrant for the system of forces. (Assume that the surface is perfectly smooth).

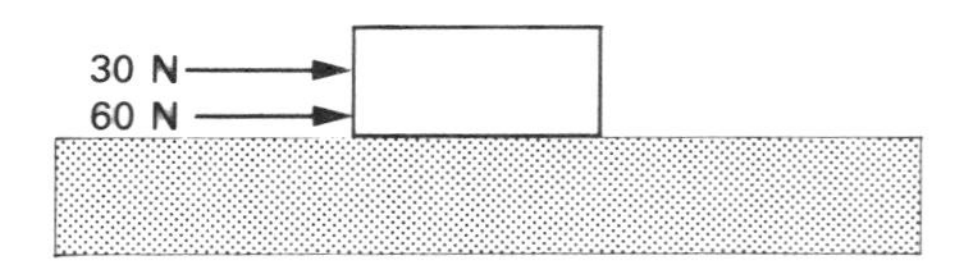

Diagram 4.14

Solution

i Because the applied forces are both horizontal to the right, the resultant will be 90 N horizontally to the right.

ii Because the combined effect of the two forces is 90 N horizontally to the right, this is what has to be balanced.

Therefore, the equilibrant = 90 N horizontally to the left, i.e. the equilibrant is equal, but opposite to the resultant. This is always true no matter how many forces there are.

When dealing with more than two applied forces we simply have more vectors on the vector diagram. The resultant is still the single straight line vector from the start of the first vector to the end of the last vector. The equilibrant is still equal, but opposite to the resultant.

Example 12

Three forces act at a point as shown in Diagram 4.15**a**. Find **i** the resultant force, and **ii** the equilibrant.

Solution

From the vector diagram (Diagram 4.15**b**) we can see that:

resultant, $R = 74$ N ↗ $7\frac{1}{2}°$
therefore the equilibrant = 74 N $7\frac{1}{2}°$ ↙

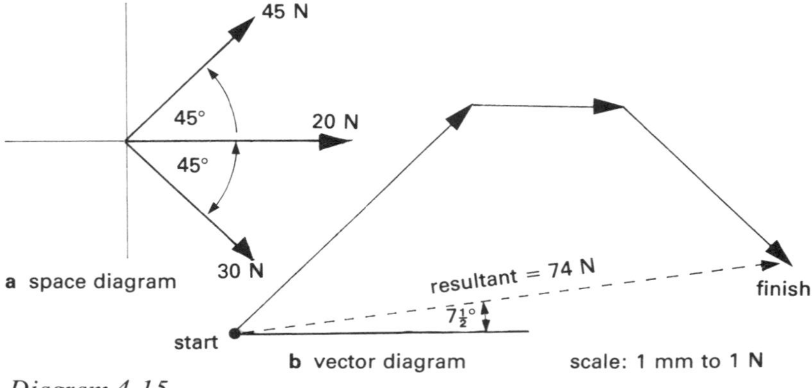

Diagram 4.15

Example 13

Four forces act on a ring as shown in Diagram 4.16**a**. Find **i** the resultant force on the ring, and **ii** the equilibrant. (The vector diagram shown in Diagram 4.16**b** is just one of the many ways that it could have been drawn.)

Solution

Therefore from the vector diagram:

i resultant, $R = 6{\cdot}7$ kN 60° ↘
ii therefore equilibrant = 6·7 kN ↖ 60°

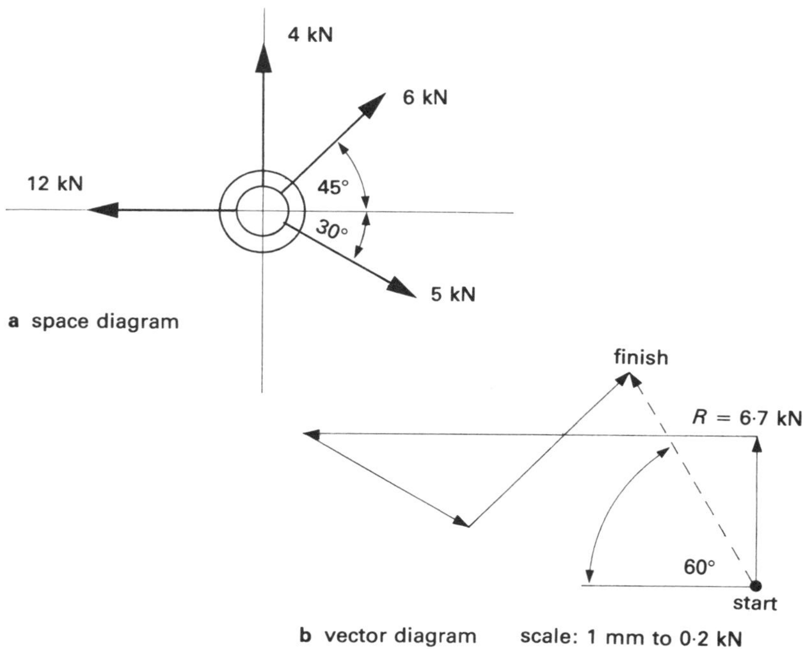

Diagram 4.16

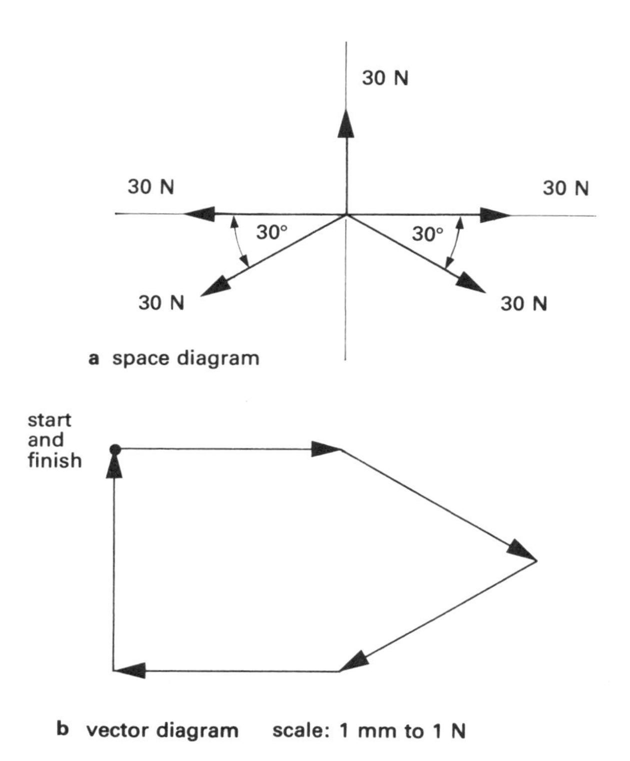

Diagram 4.17

Example 14

A system of forces is shown in Diagram 4.17**a**. Find out if the system is in equilibrium or not.

Solution

We can see, from the vector diagram in 4.17**b**, that the vectors form a closed diagram. Therefore there is no resultant force and therefore the force system is in a state of equilibrium.

Note

i For our purposes, we will assume that all forces act in one plane, i.e. co-planar forces. We will not get involved in three-dimensional force systems at O grade.

ii Whenever there are three or more forces in a system, they will always be either parallel or concurrent. Concurrent forces are forces whose lines of action all pass through a single common point called the point of concurrency. For example, look at the force systems in Examples 12 and 13.

A timber roof truss being lowered into position

4.5 Simple frame structures

Many modern houses are built using ready-made parts. One example of this is the construction of the roof. It is quite common for builders to use roof trusses of the pattern shown in Diagram 4.18.

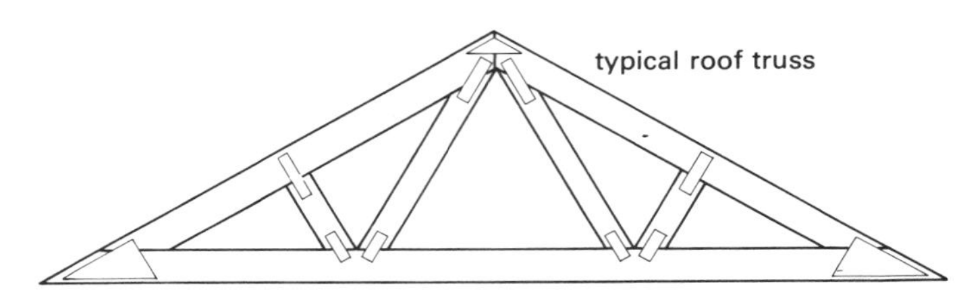

Diagram 4.18

Dockside cranes and bridges are very often made up of frameworks of steel girders. (The Forth railway bridge for example.) When frameworks are designed, the individual parts of the frame (the members) are made strong enough to support the load which each member will have to carry.

Dockside cranes

Some members will be supporting loads which tend to compress them. Others will be supporting loads tending to stretch (or extend) them. For example, the simplest frame-work of all is shown in Diagram 4.19.

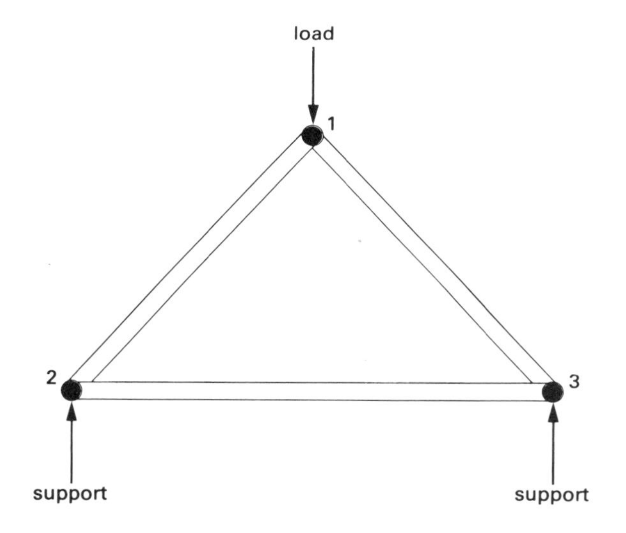

Diagram 4.19

The framework has three members pin-jointed at points 1, 2 and 3. A load acts vertically down on point 1, and the frame is held in equilibrium by supports at 2 and 3. Member 1–2 and member 1–3 are both being compressed because of the load. Member 2–3 is tending to be extended because it stops points 2 and 3 from sliding apart.

A member which is tending to be extended is in **tension** and is called a **tie**.

A member which is tending to be compressed is in **compression** and is called a **strut**.

These two kinds of members are represented as shown in Diagram 4.20. This shows what an actual member would look like and below that, how we will draw them. Therefore the simple roof truss would be drawn as in Diagram 4.21.

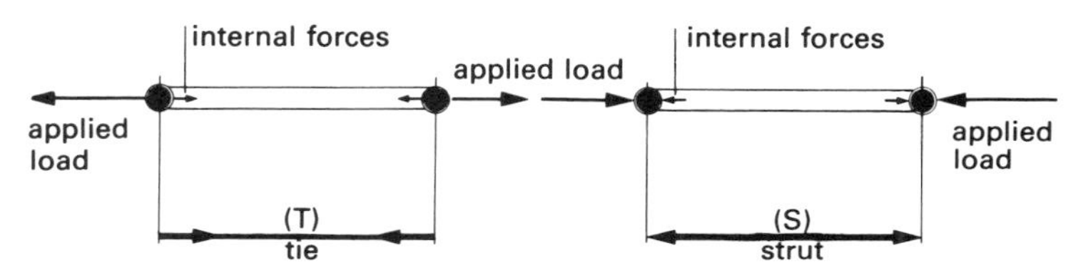

Diagram 4.20

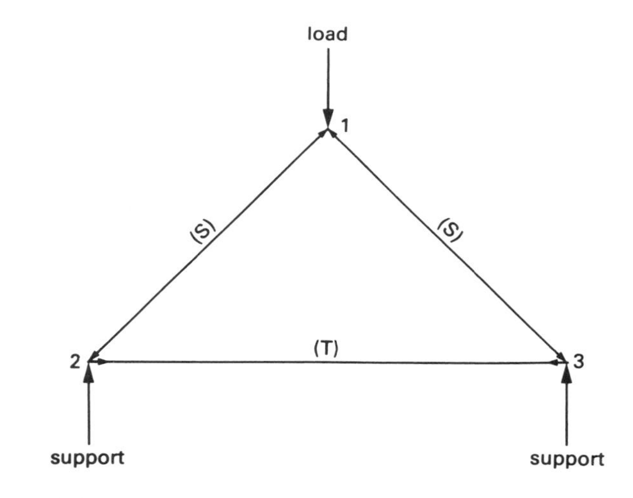

Diagram 4.21

Frames are always in equilibrium and this means that each and every joint must also be in equilibrium. (Joints are sometimes called **nodes**.)

Look at Diagram 4.21 again. There are three joints (1, 2 and 3). Each joint (or node) is a force system. In this case each has three forces – one external force and two member forces. Since each joint must be in equilibrium, the vector diagram for the forces at that joint **must** be a closed diagram.

To help us to identify each member and each force we can use a labelling system known as **Bow's Notation**. We label the spaces created by the frame; only one letter to one space. On the space diagram we use block capital letters. On the vector diagram, to name each vector, we will use small letters. For example, a roof truss as shown in Diagram 4.22, carries a vertical load of 100 kN applied at the apex. It is supported by simple supports R_1 and R_2. Because the frame is symmetrical both supports share the load equally. Therefore

$R_1 = 5$ kN and
$R_2 = 5$ kN

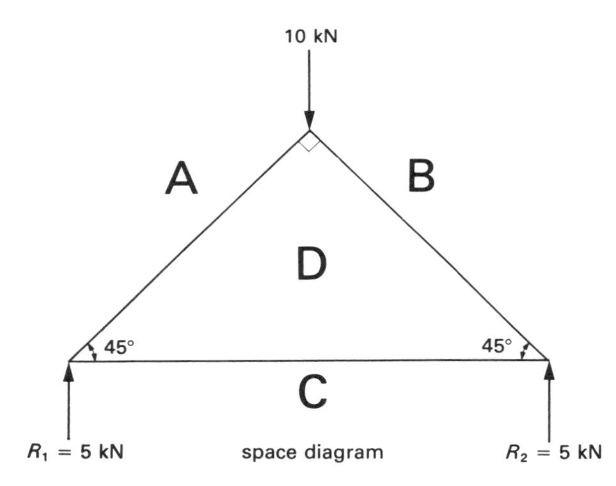

Diagram 4.22

Because there are three 'external' forces, there are three external spaces – A, B and C. Also there is only one internal space – D.

These letters are now used to name the forces and the members. The three members are AD, BD and CD. The three external forces are AB, BC and CA where AB = 10 kN, BC = R_2 = 5 kN and CA = R_1 = 5 kN.

Look at the force system at the apex. We have one applied force (10 kN) and two member forces which, as yet, are unknown. Let us draw the vector diagram for the forces at the apex to find the internal forces in the two members.

Since we know **all** the directions of all these forces **and** we know the magnitude of one of them, then we can draw the vector diagram as shown in Diagram 4.23.

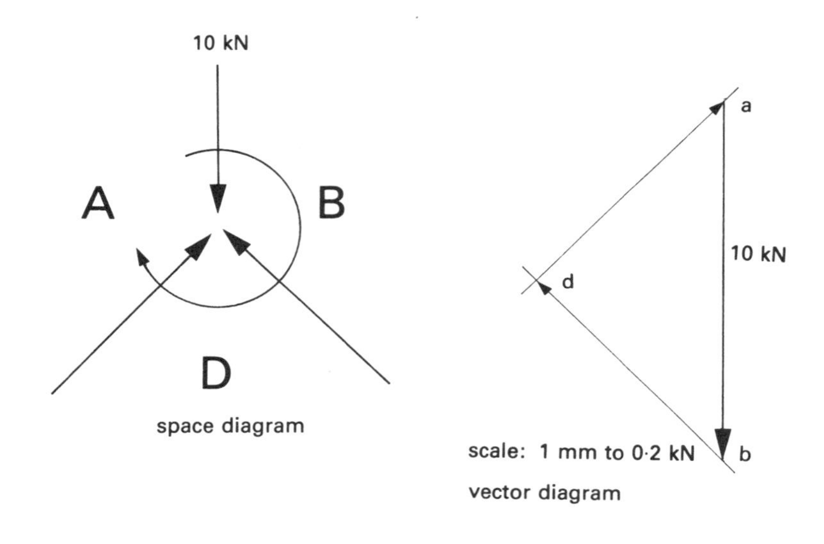

Diagram 4.23

The diagram is obtained by taking each force in a particular order – clockwise around the joint – starting with the force we know in magnitude and direction (AB). The vector for force AB is $\overrightarrow{ab}$.

The vector for member force BD **must** start from b and go in its correct direction (45°). As yet we do not know what length $\overrightarrow{bd}$ is so we draw a line which is longer than we will need. Then we take the last force, member force DA. We know that when the vector diagram is finished it must be a closed diagram. Therefore we must draw the last vector $\overrightarrow{da}$ so that it will finish at a. Again we do not know the length of the vector $\overrightarrow{da}$ but we do know its direction. So we draw a line through a at the correct angle, again longer than we will need.

These two lines for member forces BD and DA will intersect at one point on the vector diagram. This point is point d. Therefore we now have the lengths of vectors $\vec{bd}$ and $\vec{da}$, and using the scale of the diagram we can calculate the member forces.

Therefore from the vector diagram, member force BD is 7·1 kN and member force DA is 7·1 kN. Because we have a closed diagram, and the forces are in equilibrium, the vectors must follow each other 'nose to tail' from a to b, b to d and d to a. Therefore the vector diagram not only gives us the magnitude of the member forces, but it also gives us the directions of the member forces at this node. Hence members BD and DA are both struts.

Remember that the internal force in a member is the same magnitude at both ends, but opposite direction (See Diagram 4.20). This means that once we have found the member force at one joint, we know what that member force will be at the next joint it goes to. Now look at the force system for the lower right hand joint of the frame in Diagram 4.22. We can redraw this system as in Diagram 4.24.

This time we have two forces which we know in magnitude and direction. Member force DB we have just found to be 7·1 kN and its direction is 'down to the right' at this joint. Support R_2 is a force of 5 kN vertically up.

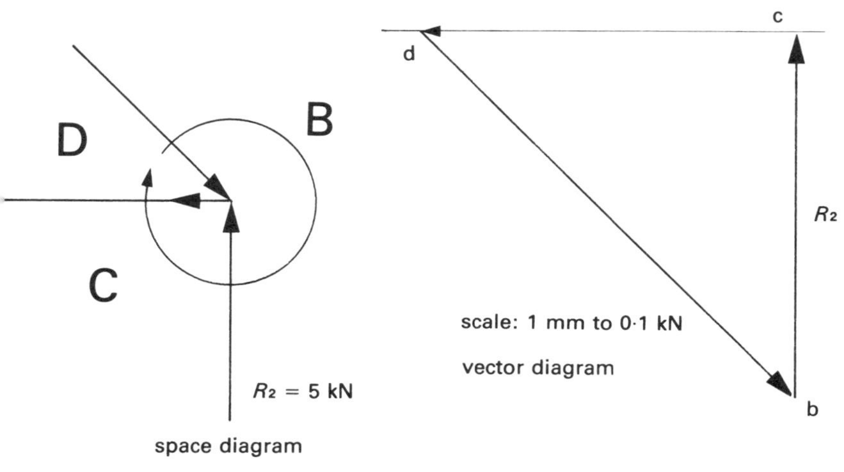

Diagram 4.24

Remember that the vector diagram is drawn by taking forces in turn from the space diagram, going clockwise round the joint, starting with the forces we know; DB then BC. The third and closing vector simply joins c to d to give $\vec{cd}$ horizontal. The same could equally well have been done for the lower left hand joint (see Diagram 4.25).

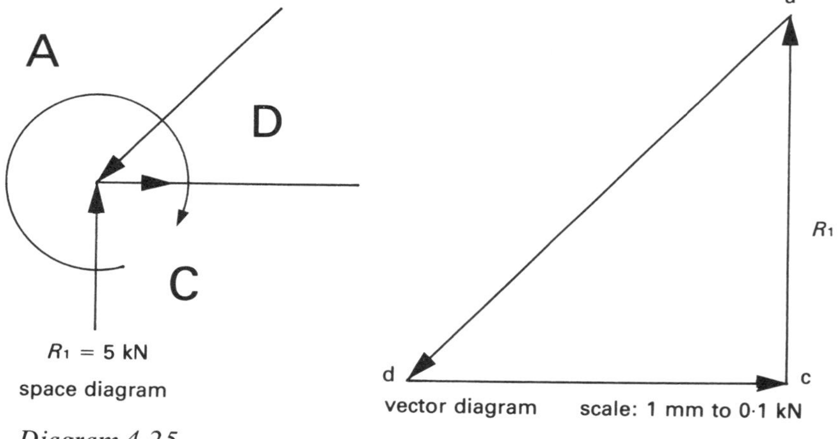

Diagram 4.25

The force in member CD is found, as before, by measuring the length of vector $\vec{cd}$ to the scale used. Therefore member force CD is 5 kN and member CD is a tie.

In practice, it is easier to use just one vector diagram for all the joints. Each time we move to another joint we keep building vectors on the same vector diagram. For this example, the combined vector diagram would be as shown in Diagram 4.26.

Note

i On a combined vector diagram we **cannot** use arrowheads.

ii To find the nature of a force in a member (i.e. whether it is a strut or a tie) we must look at one joint and read round the vector diagram for the letters round the joint in a clockwise direction.

In this case for the apex joint, the letters going clockwise are A then B then D then A again. Therefore the force directions, on the vector diagram, are from a to b, from b to d and from d to a.

iii

$\overrightarrow{ab} = 10$ kN
$\overrightarrow{bc} = R_2 = 5$ kN
$\overrightarrow{ca} = R_1 = 5$ kN

Therefore, as we expected, the supports R_1 and R_2 share the applied load equally.

Sometimes, when we cannot work out the forces of the supports to begin with, we can use the combined vector diagram to find R_1 and R_2.

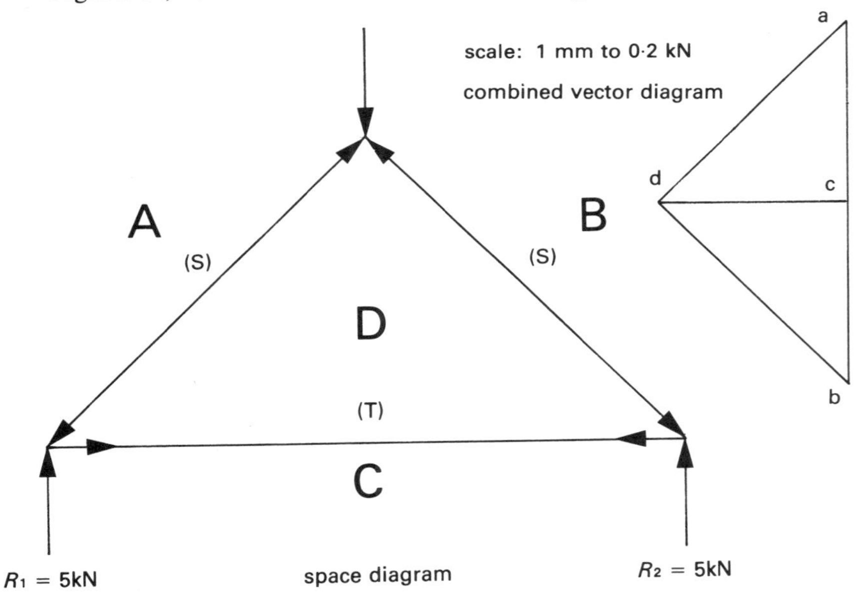

Diagram 4.26

Summary of method for frame structure

1 Always draw an accurate scale drawing of framework for a space diagram.

2 Look at the members carefully and try to decide which are struts and which are ties.

3 Work out the forces of the supports if you can at this stage.

4 Letter the space diagram using Bow's Notation.

5 Choose a joint to start at which has no more than two unknown forces.

6 Draw the vector diagram carefully, lettering the vectors. Do not use arrowheads. Make sure that the scale used gives a big enough vector diagram.

7 'Build' on the first vector diagram for all the other joints (always going round each joint clockwise).

8 When the combined vector diagram is finished and all member forces found, present the answers in tabulated form. For example:

Member	Force (in kN)	Nature
AD	7·1	Strut
BD	7·1	Strut
CD	5·0	Tie

Then state the forces in the supports, showing the direction:

$R_1 = 5{\cdot}0$ kN ↑
$R_2 = 5{\cdot}0$ kN ↑

Example 15

Find the magnitude and nature of the forces in the members of the frame structure shown in Diagram 4.27. Also find the forces of the supports, R_1 and R_2.

Solution

From the combined vector diagram:

Member	Force (in kN)	Nature
AD	5·0	Strut
BD	8·7	Strut
CD	4·3	Tie

Also $R_1 = 2{\cdot}5$ kN ↑
and $R_2 = 7{\cdot}5$ kN ↑

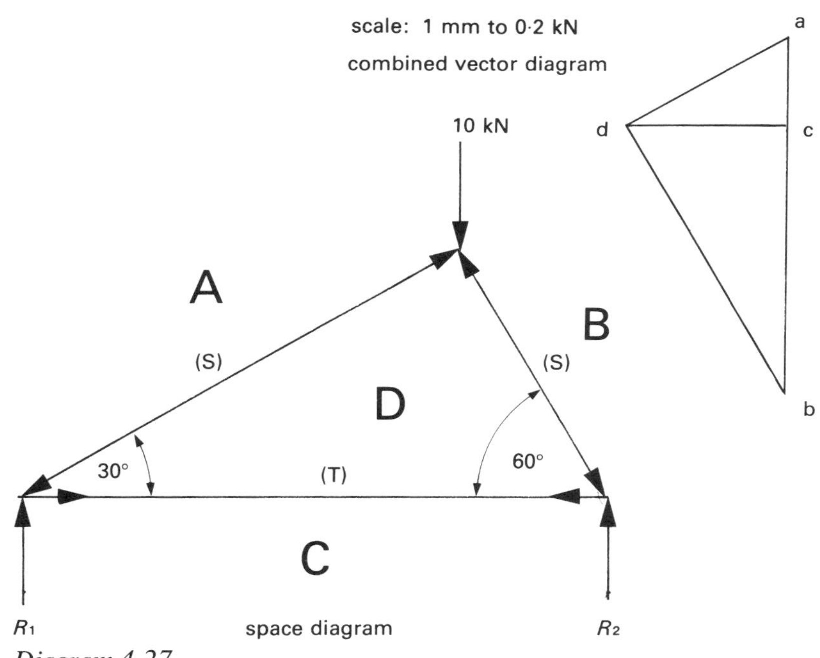

Diagram 4.27

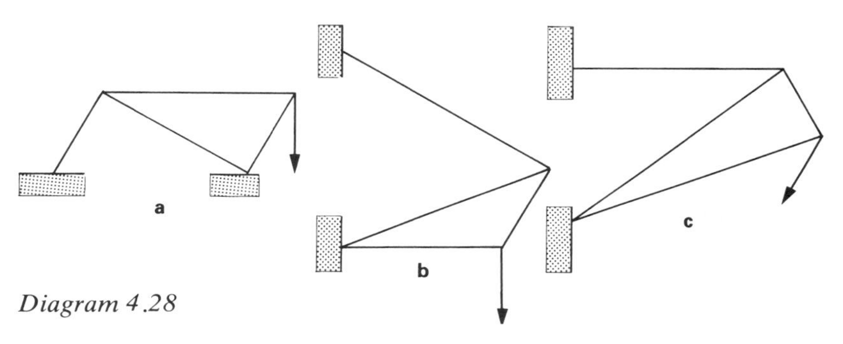

Diagram 4.28

Some possible arrangements of four members

Instead of just three members in a frame we could have four members – a triangular frame plus one external member (Diagram 4.28). For four members, the method is exactly the same as already described for the simple truss with three members.

4.6 Ways of supporting frames

Simple supports

These are knife-edge supports. They always act vertically and therefore the force of the supports, called the reactions, always act vertically (see Diagram 4.29**a**).

Single member end

When we have a four-member frame, one of the members is attached to a support point **on its own**. In this case the reaction at the support **must** be equal and opposite to the member end force, otherwise there would not be equilibrium (see Diagram 4.29**b**).

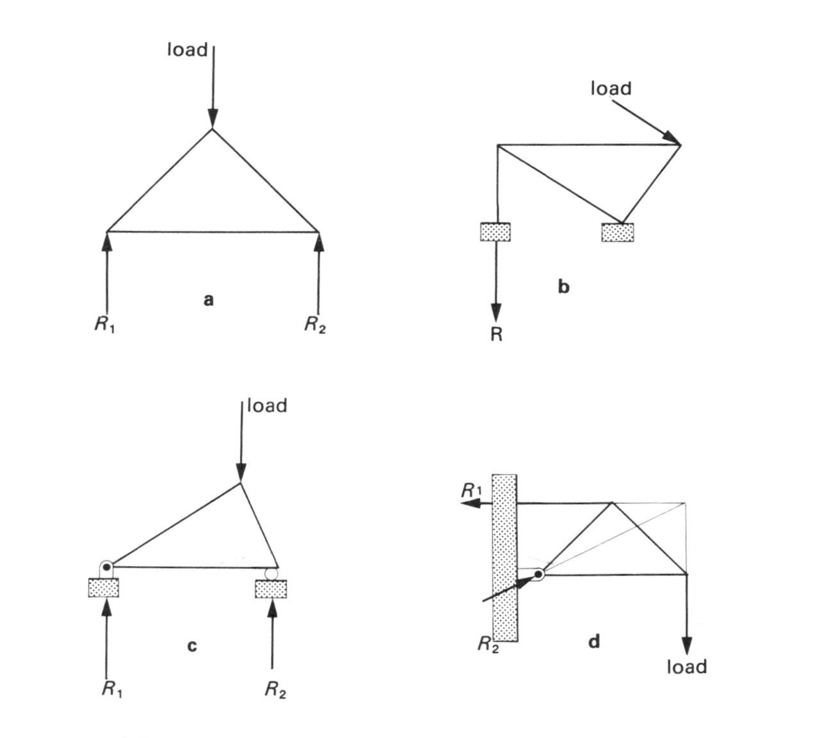

Diagram 4.29

Roller support

Whenever there are rollers, the reaction will always be outwards at right angles from the surface that the rollers are on (see Diagram 4.29**c**).

Hinge or pivot

For this kind of support the reaction can be in any direction – this would have to be found from the combined vector diagram or from the fact that for three (external) forces to be in equilibrium they must be parallel or their lines of action must be concurrent (see Diagram 4.29**c** and **d**).

Example 16

A wall-mounted crane carries a load of 8 kN as shown in Diagram 4.30. Find the magnitude and nature of the forces in the members and find the reaction forces R_1 and R_2.

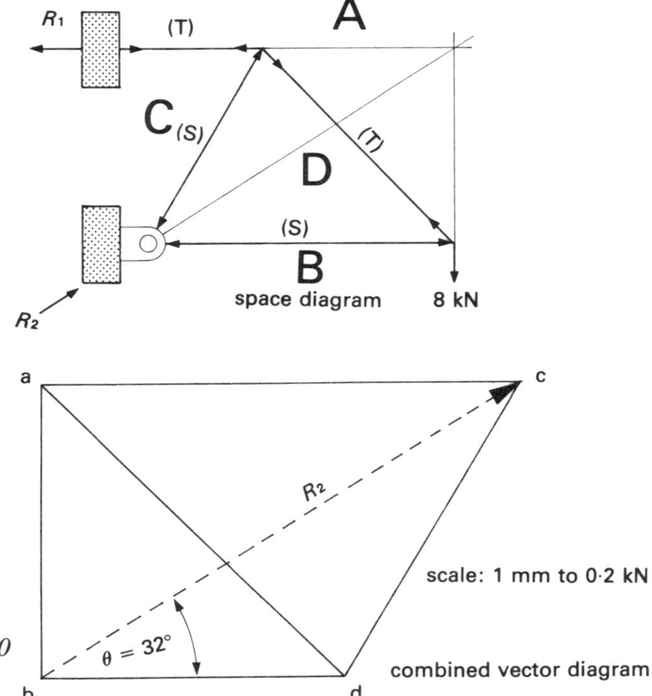

Diagram 4.30

Solution

From the combined vector diagram:

Member	Force (in kN)	Nature
AC	12·6	Tie
AD	11·3	Tie
BD	8·0	Strut
CD	9·2	Strut

Also $R_1 = 12{\cdot}6$ kN ←
and $R_2 = 14{\cdot}9$ kN ⦣ 32°

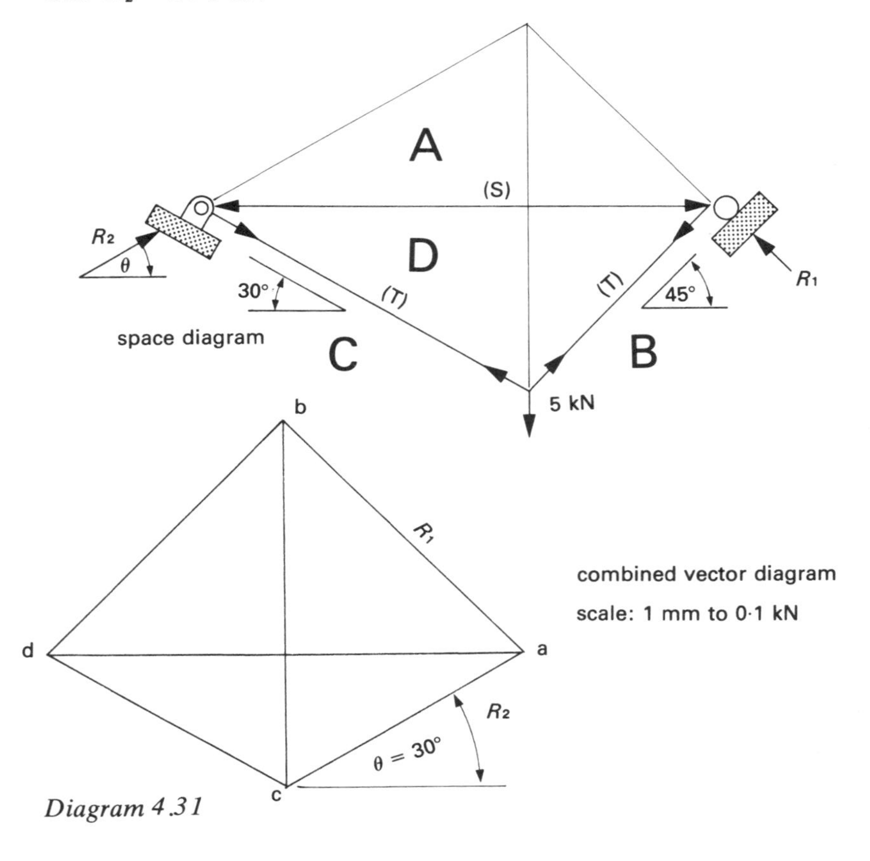

Diagram 4.31

Example 17

For the structure shown in Diagram 4.31 find **i** the reactions at the hinge and roller, and **ii** the magnitude and nature of the forces in the members.

From the combined vector diagram:

Member	Force (in kN)	Nature
AD	6·3	Strut
BD	4·5	Tie
CD	3·7	Tie

Also $R_1 = 4{\cdot}5$ kN 45°↘
and $R_2 = 3{\cdot}7$ kN ↗ 30°

4.7 Moment of a force

The **moment** of a force is the turning effect that a force has when it acts on a body which can pivot or rotate. For example, the force at the end of a spanner makes the spanner turn the nut (Diagram 4.32).

For a very tight nut, we would probably use a longer spanner to get a bigger moment (turning effect). If there wasn't a longer spanner we would have to try a greater force to get a bigger moment. So we actually know from experience that there is a connection between the force, the distance to the turning point (or pivot) and the moment or turning effect.

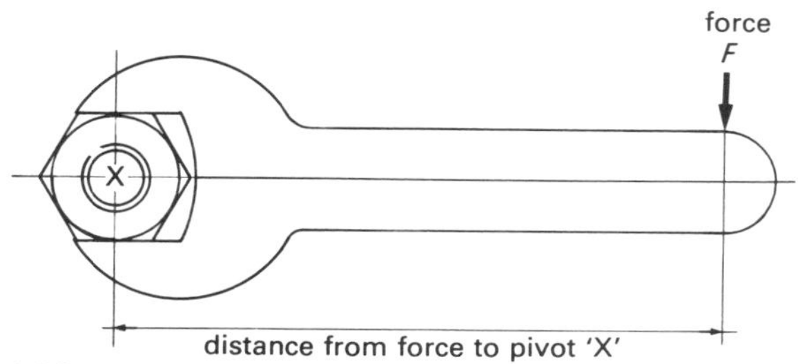

Diagram 4.32

Perhaps a less obvious example of moments is a see-saw. For a big person and a small person to make a see-saw work, the big person has to be nearer to the pivot. This is so that the moments at each side of the pivot balance each other.

The value of a moment is found by multiplying the force by the distance to the pivot.

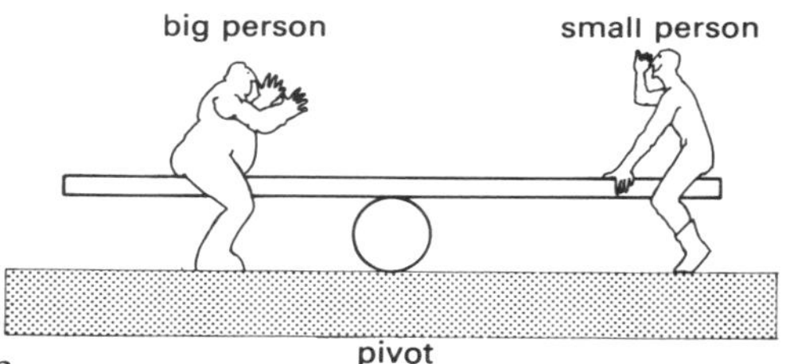

Diagram 4.33

Note

The distance to the pivot is the shortest distance from the line of action of the force to the pivot, i.e. the perpendicular distance. Therefore:

moment of force = force (F) × perpendicular distance to the pivot (x)
moment = Fx

The units for measuring a moment must therefore be newton-metres (N m). We can summarise all these things in one definition.

The **moment** of a force is the turning effect of the force. It is found by multiplying the force by the perpendicular distance from the line of action of the force to the pivot point and its units are newton-metres.

Example 18

Calculate the moment of the force about O for each of the arrangements shown in Diagram 4.34.

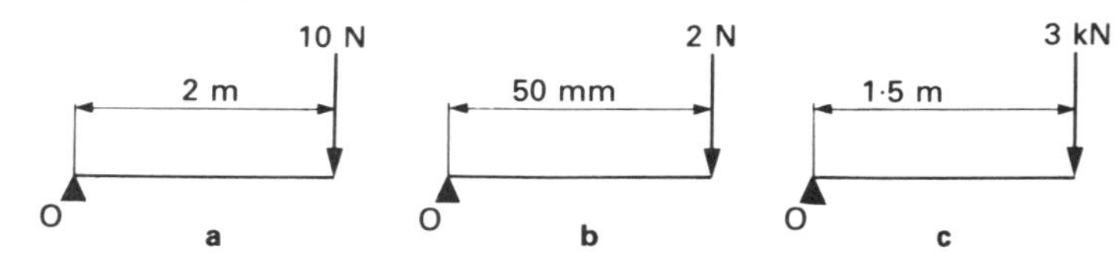

Diagram 4.34

Solution

a Moment $= Fx$
$= 10 \times 2$
$= 20$ N m

b Moment $= Fx$
$= 2 \times 50$
$= 100$ N mm

c Moment $= Fx$
$= 3 \times 1{\cdot}5$
$= 4{\cdot}5$ kN m

When there are two or more forces acting at one side of a pivot, we find the total moment by adding the two individual moments together. We can only do this if the units are the same so we must never mix units (e.g. N m and kN m don't mix).

Example 19

Calculate the total moment about O for each of the arrangements shown in Diagram 4.35.

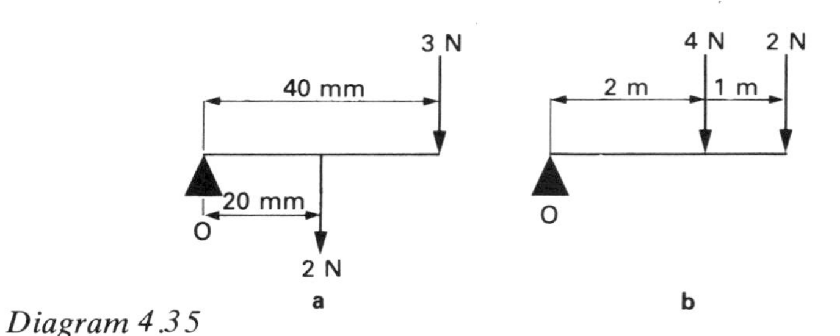

Diagram 4.35

Solution

Taking moments about point O:

a Total moment $= F_1 x_1 + F_2 x_2$
$= (2 \times 20) + (3 \times 40)$
$= 40 + 120$
$= 160$ N mm

b Total moment $= F_1 x_1 + F_2 x_2$
$= (4 \times 2) + (2 \times 3)$
$= 8 + 6$
$= 14$ N m

However, for beams or levers which are in equilibrium (i.e. balanced) the total moment on one side of the pivot must equal the total moment on the other side of the pivot (remember the see-saw). So that we can identify which side of the pivot we are dealing with, we will talk about 'clockwise moments' and 'anticlockwise moments'. **The Principle of Moments** states that for a body to be in equilibrium the sum of the clockwise moments must equal the sum of the anticlockwise moments about a point in the plane.

Example 20

The beams shown in Diagram 4.36 are all in equilibrium. Find the unknown quantity for each arrangement.

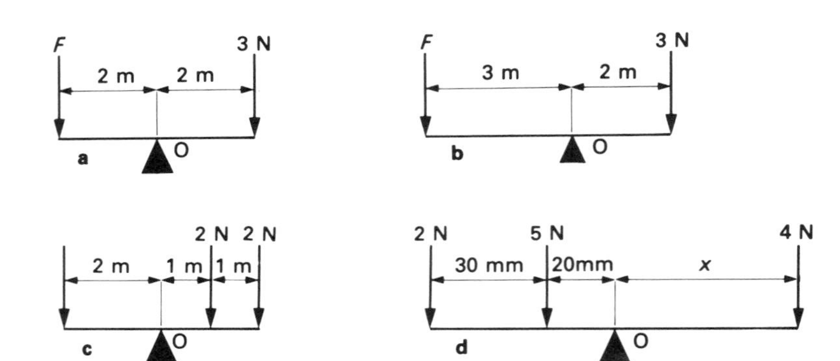

Diagram 4.36

Solution

a For equilibrium, taking moments about O:

sum of anticlockwise moments = sum of clockwise moments

therefore $F \times 2 = 3 \times 2$

therefore $F = 3$ N

b For equilibrium, taking moments about O:

sum of anticlockwise moments = sum of clockwise moments

therefore $F \times 3 = 3 \times 2$

$= 6$

therefore $F = 2$ N

c For equilibrium, taking moments about O:

sum of anticlockwise moments = sum of clockwise moments

therefore $F \times 2 = (2 \times 1) + (2 \times 2)$

$= 2 + 4$

$= 6$

therefore $F = 3$ N

d For equilibrium, taking moments about O:

sum of clockwise moments = sum of anticlockwise moments

therefore $4x = (5 \times 20) + (2 \times 50)$

$= 100 + 100$

$= 200$

therefore $x = 50$ mm

So far all the examples have dealt with horizontal see-saw type beams or levers, but the Principle of Moments applies to all kinds of arrangements: vertical poles with guy ropes, wheelbarrows, scissors, even opening doors. Also, the forces we have dealt with have all been vertical, but they need not have been.

Example 21

A variety of arrangements, involving the Principle of Moments, are shown in Diagram 4.37. Find the unknown quantities in each case to give a state of equilibrium.

Solution

a Since we are dealing with forces, we must first find the gravitational force acting on the 50 kg mass. Therefore:

$F_g = 50 \times 9{\cdot}81 = 491$ N

For equilibrium, taking moments about O:

sum of anticlockwise moments = sum of clockwise moments

therefore $F \times 0{\cdot}8 = 491 \times 0{\cdot}3$

$F = 184$ N ↑

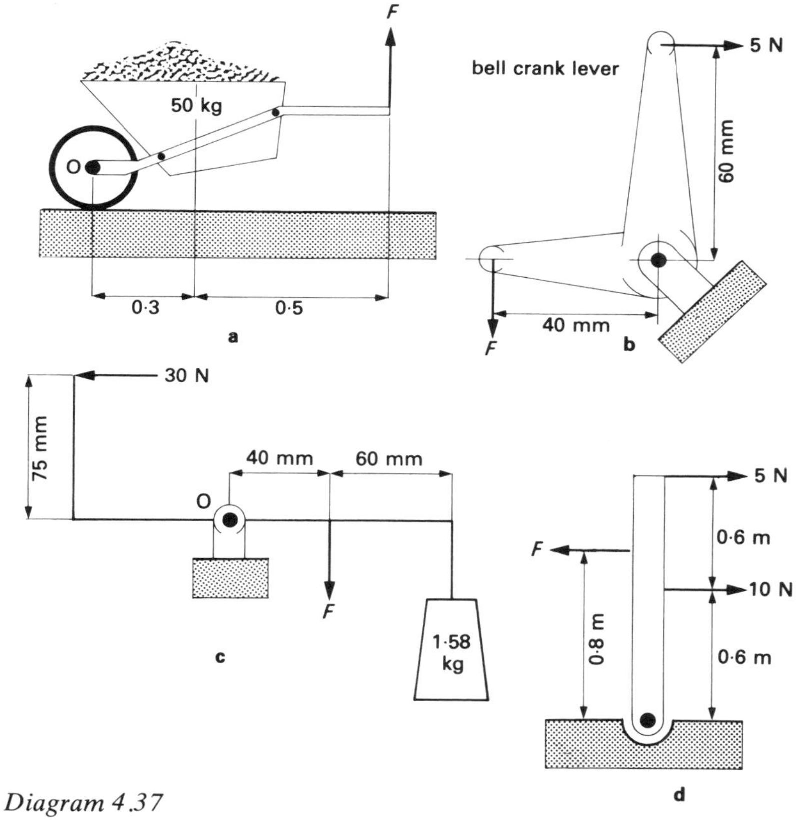

Diagram 4.37

b For equilibrium, taking moments about O:

sum of anticlockwise moments = sum of clockwise moments

therefore $F \times 40 = 5 \times 60$

$40\,F = 300$

$F = 7{\cdot}5$ N ↓

c We must find the gravitational force on the mass of 1·58 kg.

$F_g = 1{\cdot}58 \times 9{\cdot}81 = 15{\cdot}5$ N

For equilibrium, taking moments about O:

$$\begin{aligned}\text{sum of clockwise moments} &= \text{sum of anticlockwise moments}\\ \text{therefore } (F \times 40) + (15{\cdot}5 \times 100) &= (30 \times 75)\\ 40F + 1550 &= 2250\\ 40F &= 2250 - 1550\\ 40F &= 700\\ F &= 17{\cdot}5 \text{ N} \downarrow\end{aligned}$$

d For equilibrium, taking moments about O:

$$\begin{aligned}\text{sum of anticlockwise moments} &= \text{sum of clockwise moments}\\ \text{therefore } (F \times 0{\cdot}8) &= (10 \times 0{\cdot}6) + (5 \times 1{\cdot}2)\\ 0{\cdot}8F &= 6 + 6\\ F &= \frac{12}{0{\cdot}8}\\ F &= 15 \text{ N} \leftarrow\end{aligned}$$

For levers or beams which have inclined forces acting on them, we must be very careful to work out the **correct** distance between the line of action of the force and the pivot point. (Remember that the correct distance is the perpendicular distance.)

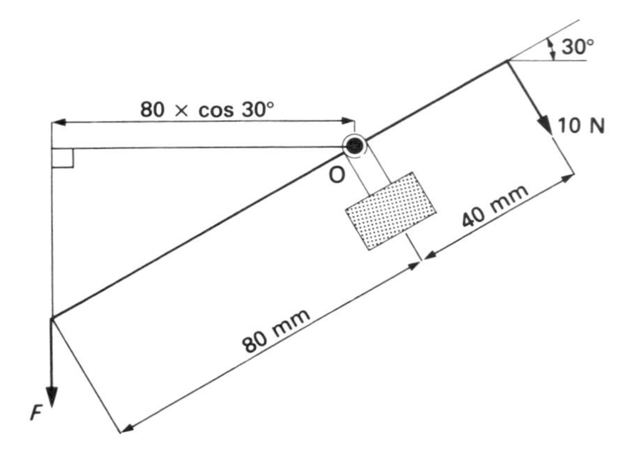

Diagram 4.38

Example 22

The beam shown in Diagram 4.38 is in equilibrium under the action of the forces indicated. Find the magnitude of F.

Solution

The perpendicular distance between the line of action of F and the pivot point O is 80 cos 30 mm. Therefore for equilibrium, taking moments about O:

$$\begin{aligned}\text{sum of anticlockwise moments} &= \text{sum of clockwise moments}\\ \text{therefore } (F \times 80 \cos 30) &= (10 \times 40)\\ F \times 69 &= 400\\ F &= 5{\cdot}8 \text{ N} \downarrow\end{aligned}$$

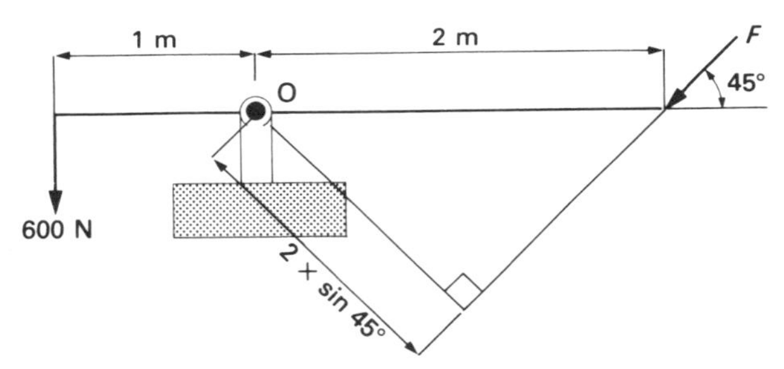

Diagram 4.39

Example 23

Find the magnitude of F so that the beam shown in Diagram 4.39 will be in equilibrium.

Solution

The perpendicular distance between the line of action of F and the pivot O is 2sin 45 m. Therefore for equilibrium, taking moments about O:

$$\begin{aligned}\text{sum of clockwise moments} &= \text{sum of anticlockwise moments}\\ \text{therefore } (F \times 2\sin 45) &= (600 \times 1)\\ F &= \frac{600}{2 \times 0{\cdot}707}\\ F &= 423 \text{ N} \quad 45^\circ \swarrow\end{aligned}$$

Note

Examples 22 and 23 have been worked out using trigonometry to find the perpendicular distance required, but accurate scale drawings of the arrangement can be used and the perpendicular distance can be drawn and measured to scale.

General conditions of equilibrium

In our work on Statics, we have seen that a body will be in equilibrium when the vector diagram of the forces forms a closed diagram, i.e. there is no resultant force. Also, for equilibrium there must be no resultant moment.

No resultant force

This condition of equilibrium can be dealt with in two ways.

a *Graphically.* The vector diagram must close.

b *Analytically (by calculation).* The algebraic sum of the forces in the x direction must be zero, and the algebraic sum of the forces in the y direction must be zero. The x and y directions are usually taken as horizontal and vertical, but they can be any pair of directions which are at right angles to each other. This way of dealing with force equilibrium is most useful when we resolve sloping forces into their rectangular components (see Section 4.3).

No resultant moment

As we have already seen, the Principle of Moments must apply to any body in equilibrium. If we call 'clockwise' the positive direction for moments, then 'anticlockwise' will be the negative direction and we can say that the algebraic sum of the moments must be zero.

In mathematics, the symbol which means 'the algebraic sum of' something is the Greek letter Σ (sigma). Therefore, the General Conditions of Equilibrium can be written in a shorthand form as:

1 The vector diagram of the forces must close
or $\Sigma F_x = 0$ and $\Sigma F_y = 0$

2 Sum of the clockwise moments = sum of the anticlockwise moments
or $\Sigma M = 0$

Beam reactions

Normally we use the word 'lever' to mean a bar or rod that has only one supporting point or pivot point. A **beam** is usually supported at two points and there are two main ways of doing this.

1 Simple supports (knife-edges) are used when there is no sideways tendency to move the beam. For example, the loaded beam shown in Diagram 4.40.

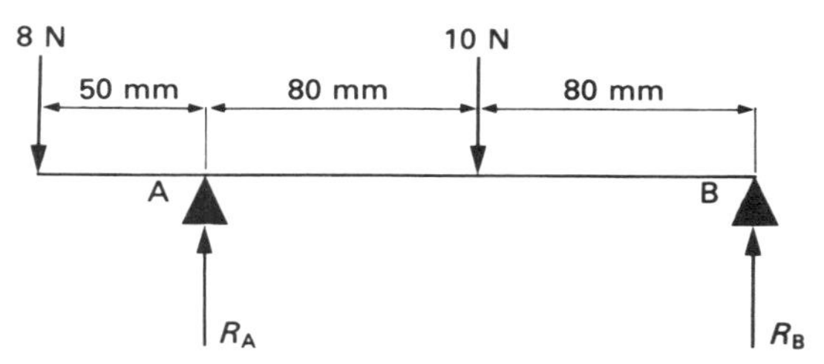

Diagram 4.40

There are two unknown forces – the reaction at A and the reaction at B. However, we do know that the beam is in equilibrium and that we can use the General Conditions of Equilibrium. When we take moments, we must be sure to take one of the support points as a pivot point. This means that in our moments equation there will be only one unknown force; the reaction at the other pivot.

For this example, let us first take moments about point A; therefore for equilibrium:

sum of the anticlockwise moments = sum of the clockwise moments

$$\text{therefore } (R_B \times 160) + (8 \times 50) = (10 \times 80)$$
$$160\,R_B = 800 - 400$$
$$R_B = \frac{400}{160}$$
$$R_B = 2{\cdot}5\text{ N} \uparrow$$

Now let us take moments about point B, so that we can find R_A; therefore for equilibrium:

sum of the clockwise moments = sum of the anticlockwise moments

therefore
$$(R_A \times 160) = (10 \times 80) + (8 \times 210)$$
$$160\ R_A = 800 + 1680$$
$$R_A = \frac{2480}{160}$$
$$R_A = 15{\cdot}5\ \text{N}\ \uparrow$$

We can check that the result makes sense by remembering that $\Sigma F_y = 0$ for equilibrium.

Therefore sum of upward forces = sum of downward forces
i.e. $(R_A + R_B)$ should equal (10N + 8N)
and 15·5 + 2·5 does equal 18
and so the answers make sense.

2 Hinge and roller supports are used when a beam has sideways loading because the hinge can react to this kind of loading whereas knife-edges cannot. We know that the reaction at a roller support is always at right angles to the surface on which the roller rests, but the reaction at a hinge can be in any direction.

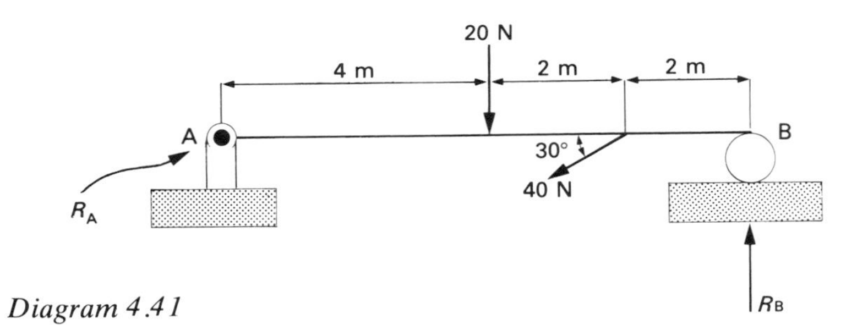

Diagram 4.41

Example

Examine the loaded beam shown in Diagram 4.41. There are **three** unknown quantities here; the magnitude of R_B, the magnitude of R_A and the direction of R_A. We can solve this problem in at least two ways using the general conditions of equilibrium:

i Take moments about A and hence find R_B. Then draw the vector diagram – the 'closer' is R_A in magnitude and direction.

or **ii** Take moments about A and hence find R_B. Then resolve the 40N force into its vertical and horizontal components. Also consider R_A to be made up of its two components (one vertical, the other horizontal). Apply the conditions for force balance, $\Sigma F_x = 0$ and $\Sigma F_y = 0$. This will let us find the two components of R_A and therefore we can find force R_A in magnitude and direction.

Solution **i**

For equilibrium, taking moments about A:

sum of anticlockwise moments = sum of clockwise moments
therefore
$$(R_B \times 8) = (20 \times 4) + (40 \times 6 \sin 30)$$
$$8\ R_B = 80 + 120$$
$$R_B = \frac{200}{8}$$
$$\therefore R_B = 25\ \text{N}\ \uparrow$$

Now that we know three of the four forces in magnitude and direction, we can draw the vector diagram for the forces and the 'closer' will be R_A in magnitude and direction (Diagram 4.42). From the Diagram, measuring to scale
$R_A = 37{\cdot}5$ N $\angle\ 23\tfrac{1}{2}°$

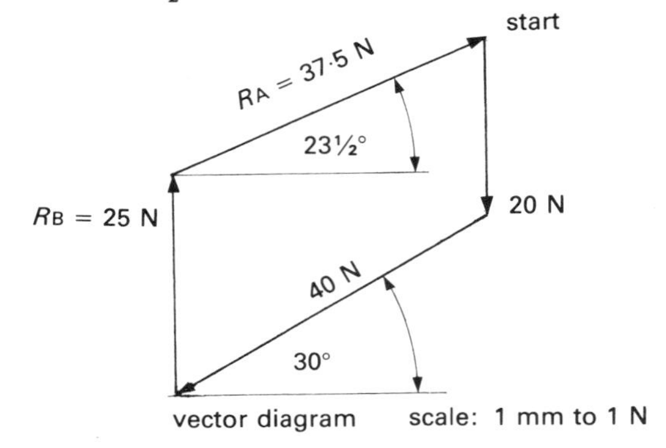

Diagram 4.42

Solution **ii**

As in **i** by taking moments about A,

$R_B = 25$ N ↑

Now, we resolve the 40 N force into its two components (Diagram 4.43).
From the diagram, measured to scale

horizontal component = 34·5 N ←
vertical component = 20 N ↓

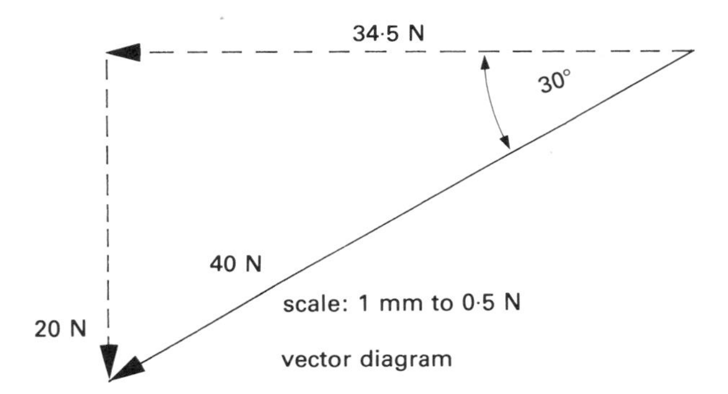

Diagram 4.43

Let the components of R_A be V and H for vertical and horizontal components. We could re-draw the original force system with the inclined forces replaced by their components (Diagram 4.44).

For equilibrium, $\Sigma F_x = 0$
therefore $H = 34{\cdot}5$ N →
also $\Sigma F_y = 0$
therefore $V + 25 = 20 + 20$
$V = 15$ N ↑

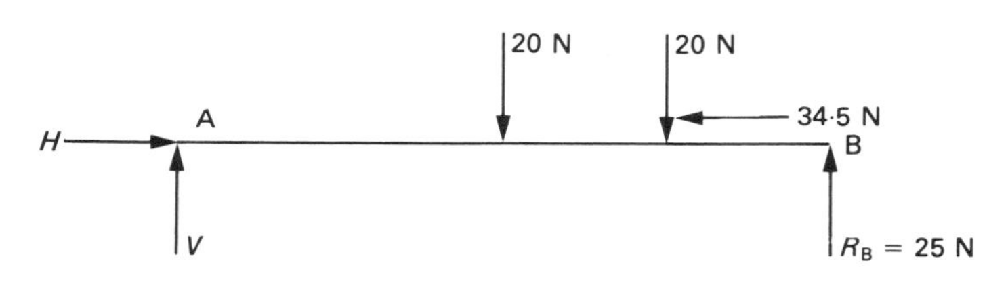

Diagram 4.44

To find R_A we must now combine its two components V and H (Diagram 4.45). Therefore, from the diagram, measured to scale:

$R_A = 37{\cdot}5$ N ⦣ $23\frac{1}{2}°$

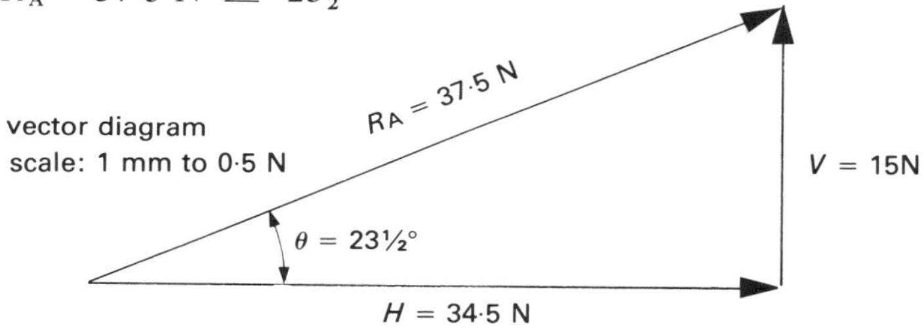

vector diagram
scale: 1 mm to 0·5 N
$R_A = 37{\cdot}5$ N
$V = 15$N
$\theta = 23\frac{1}{2}°$
$H = 34{\cdot}5$ N

Diagram 4.45

There are other ways of solving the problem, but solution **i** is the quickest and neatest way.

Torque

Torque is quite similar to a moment. When a force twists, or tries to twist a shaft, then the shaft is under the action of a torque. Usually we refer to torque when we are dealing with rotating shafts. Torque is measured in exactly the same way as a moment. It is the product of force and distance to pivot point, but for a rotating shaft the distance to the pivot point is the distance to the axis of rotation, i.e. radius.

Torque = force × radius

The units for torque must therefore be exactly the same as the units for a moment (N m etc.).

Example 24

A force of 25 N acts on the rim of a pulley which is 80 mm in diameter. Find the torque (Diagram 4.46).

Solution

Torque = force × radius
= 25 × 40
= 1000 N mm
therefore torque = 1000 N mm (or 1 N m)

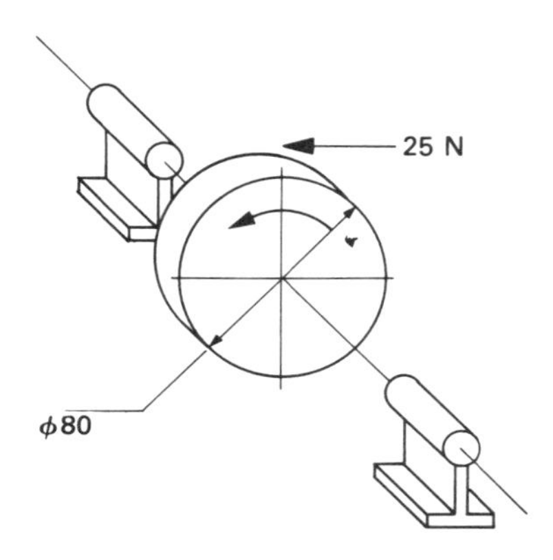

Diagram 4.46

Example 25

A shaft supplies a torque of 40 N m to a pulley which is 320 mm in diameter. What force will be supplied at the rim of the pulley?

Solution

Torque = force × radius

However, since torque is given in N m, we must make sure that the radius is in m.

Therefore $40 = F \times 0{\cdot}16$

$$F = \frac{40}{0{\cdot}16}$$

rim force, $F = 250$ N

4.8 Centroid and centre of gravity

So far we have dealt with bodies which have mass and therefore have a force of gravity acting on them. However, we have not mentioned size or shape.

All bodies are made up of small particles of matter and each particle has its own mass with its own force of gravity acting on it. For example, we can think of a metre-stick as being made up of a number of small lengths each with its own force of gravity. The total force of gravity

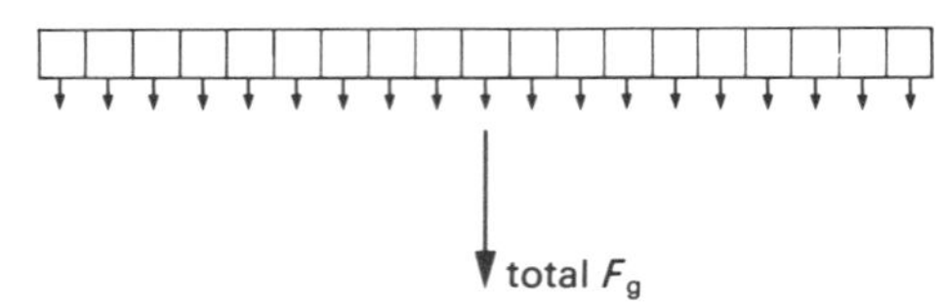

Diagram 4.47

on the metre-stick is the sum of all the individual forces of gravity (Diagram 4.47).

We know that to balance something like a metre-stick horizontally on just one support, then we would have to put the exact midpoint of the metre-stick above the support. (This is the same as balancing a ruler on one finger.)

That one particular point in the metre-stick where it can be balanced, is called the **centre of gravity** of the metre-stick.

The centre of gravity of a body is that one particular point where the total force of gravity of the body will be taken to act. For some bodies, the centre of gravity will be within the material of the body and for others it will actually be outside the body. For example, the centre of gravity of a brick is somewhere in the middle of the brick, but the centre of gravity of a hollow ring is in the space inside the ring.

Centroids

When we are dealing with objects which are of uniform thickness then we need to consider only its geometrical shape: the shape of its surface area. The centre of gravity of the object will be at the same point as the centre of area of the surface (but half way through the thickness).

The centre of area of a shape is called the **centroid**. For any geometrical shape we can find the centroid by imagining that the shape is made up of regular, very narrow strips. Each strip has its own centroid at the mid-point (like the balancing metre-stick). When each strip has its own centroid marked, we can join the centroids with a line and the overall centroid must be somewhere along this line. By imagining the shape to be made up of strips again, but this time in a different direction across the shape, we will get another line on which the overall centroid must lie.

The point where these two lines cross is the centroid of the shape. This can be shown for some common geometrical shapes.

Rectangle

Diagrams 4.48**a** and **b** show two ways of taking strips, each with its line drawn through the centroids of the strips. Diagram 4.48**c** shows the shape with both of these lines drawn on it. The point where they cross is the centroid. The centroid of a rectangle is also at the point where the diagonals intersect.

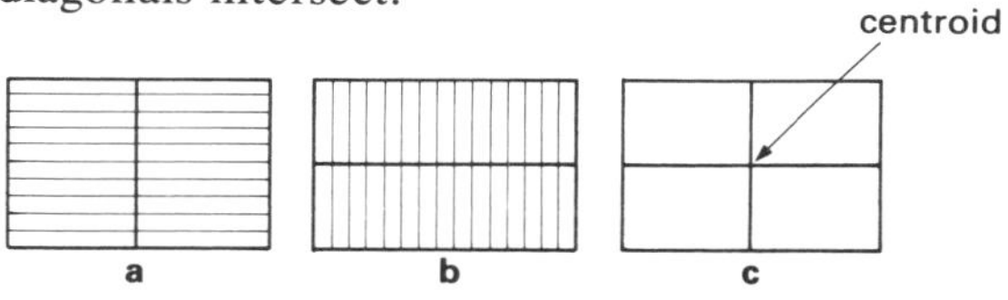

Diagram 4.48

Triangle

Diagram 4.49 shows the strips and lines for a triangle. In a triangle these lines are called **medians**. A median is a line which joins the mid-point of a side to the opposite vertex. The medians of any triangle always intersect at a point one third of the way along each median.

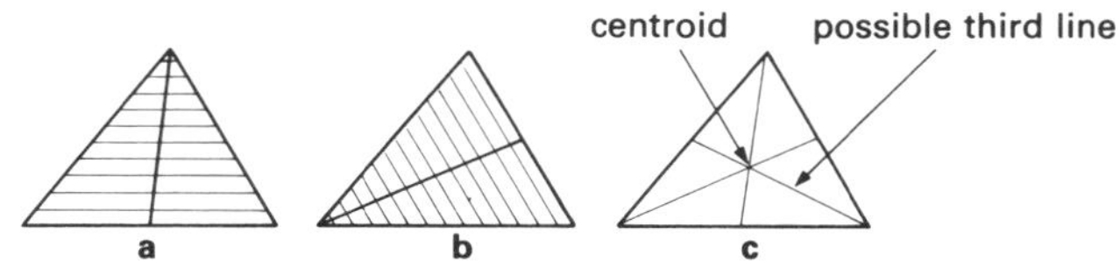

Diagram 4.49

Circle

The centroid of a circle must obviously be the centre of the circle.

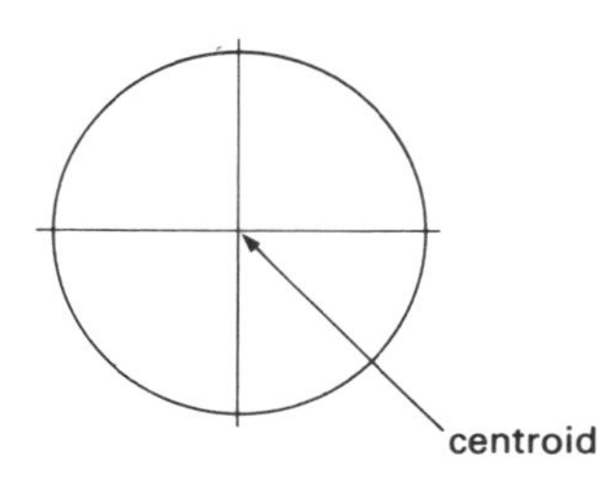

Diagram 4.50

Composite shapes

Finding the centroid of geometrical shapes, taken one at a time, is quite straightforward, but most real objects are made up of two or more geometrical shapes. Consider the shape drawn in Diagram 4.51.

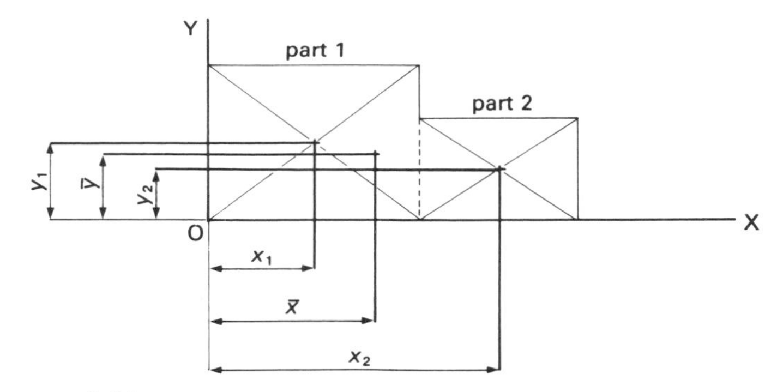

Diagram 4.51

We can think of this shape as being made up of two rectangles in two ways. It could be a large rectangle with a smaller rectangle joined on, or it could be a larger rectangle with a rectangular corner cut out. Let us take it as two rectangles joined together: rectangle 1 of area A_1 and rectangle 2 of area A_2. To find the overall centroid we will have to take area-moments about two convenient axes, say OX and OY as shown.

The area-moment for the whole shape must be the same as the sum of the individual area-moments.

We know that for each rectangle its centroid is at the intersection of its diagonals (i.e. at the point which lies half way along the central axis for that rectangle).

Let the centroid for rectangle 1 be the point (x_1,y_1) and for rectangle 2 let the centroid be the point (x_2,y_2).

Also, for the whole shape let us call the total area ΣA (i.e. the sum of all the individual areas) and let us assume that the overall centroid lies at the unknown point $(\bar{x},\bar{y})$.

To find $\bar{x}$ first, let us take area-moments about axis OY:

$$(\Sigma A)\bar{x} = A_1x_1 + A_2x_2.$$

Let us call $\Sigma(Ax)$ the 'sum of the individual area-moments'.

Therefore $\Sigma(Ax) = A_1x_1 + A_2x_2$

and $(\Sigma A)\bar{x} = \Sigma(Ax)$

or $\bar{x} = \dfrac{\Sigma(Ax)}{\Sigma A}$

In words:

$$\text{distance to centroid} = \frac{\text{sum of individual area-moments}}{\text{sum of individual areas}}$$

Also, to find $\bar{y}$, taking moments this time about axis OX, we can show that:

$$\bar{y} = \frac{\Sigma(Ay)}{\Sigma A}$$

Example 26

Find the centroid of the template shown in Diagram 4.52.

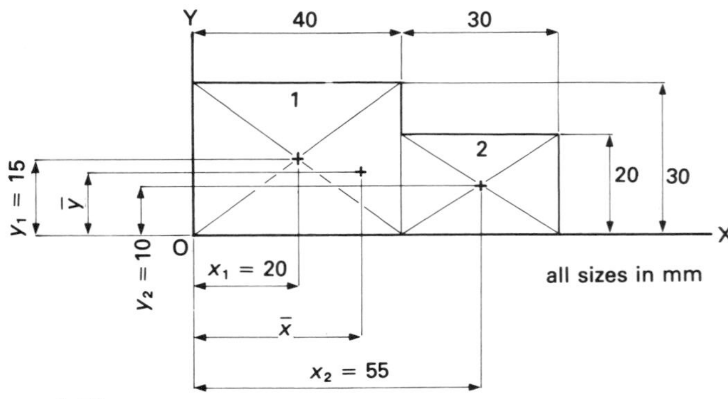

Diagram 4.52

Solution

To help collect all the necessary information it is best to use a table as shown below.

Part	Area(A)	Distance to centroid (x)	Area-moment (Ax)	Distance to centroid (y)	Area-moment (Ay)
Rectangle 1	1200 mm²	20 mm	24 000	15 mm	18 000
Rectangle 2	600 mm²	55 mm	33 000	10 mm	6000
Total	ΣA = 1800 mm²	$\bar{x}$	ΣA_x = 57 000	$\bar{y}$	ΣA_y = 24 000

Therefore $\bar{x} = \dfrac{\Sigma(Ax)}{\Sigma A} = \dfrac{57\,000}{1800} = 31{\cdot}7$ mm from OY

and $\bar{y} = \dfrac{\Sigma(Ay)}{\Sigma A} = \dfrac{24\,000}{1800} = 13{\cdot}3$ mm from OX

therefore the centroid is the point (31·7, 13·3).

Example 27

Find the centroid of the shape shown in Diagram 4.53.

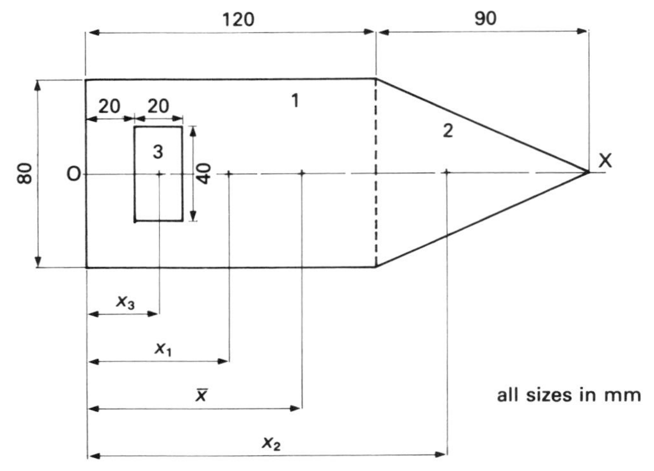

Diagram 4.53

Solution

Draw up a table as shown below. Note that because part 3 is a hole, we will give it a negative area and therefore it will have a negative area moment.

Part	Area (A) mm²	Distance to centroid from O (x)	Area-moment (Ax)
1	9600 mm²	60	576×10^3
2	3600 mm²	120 + 30 = 150	540×10^3
3	−800 mm²	20 + 10 = 30	-24×10^3
Total	ΣA = 12400 mm²	$\bar{x}$	$\Sigma Ax = 1092 \times 10^3$

Therefore $\bar{x} = \dfrac{\Sigma Ax}{\Sigma A}$

$= \dfrac{1092 \times 10^3}{12{\cdot}4 \times 10^3}$

$= 88{\cdot}1$ mm

therefore the centroid is 88·1 mm from O along OX

When dealing with bodies made up of shapes with different thickness, we must consider volumes instead of just surface areas. In such cases we will have to find volumes and volume-moments. Our equation to find the position of the centre of gravity will become:

$$\bar{x} = \frac{\Sigma(Vx)}{\Sigma V}$$

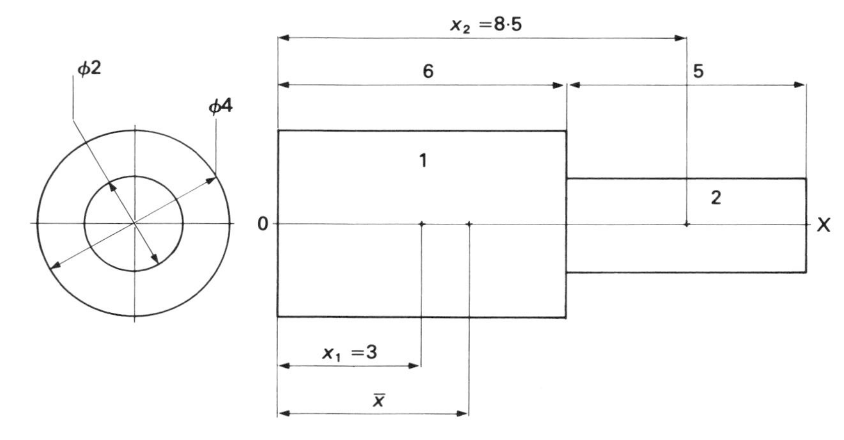

Diagram 4.54

Example 28

Find the centre of gravity for the shaft shown in Diagram 4.54 (sizes are given simply as units).

Solution

Taking volume-moments about O:

Part	Volume (V)	x	Vx
1	$(\pi \times 2^2 \times 6) = 75{\cdot}4$	3	226
2	$(\pi \times 1^2 \times 5) = 15{\cdot}7$	8·5	133
Total	$\Sigma V = 91{\cdot}1$	$\bar{x}$	$\Sigma Vx = 359$

Therefore $\bar{x} = \dfrac{\Sigma Vx}{\Sigma V}$

$= \dfrac{359}{91{\cdot}1}$

$= 3{\cdot}94$

therefore the centre of gravity is 3·94 units from O along OX.

When dealing with bodies made up of shapes of different thickness and different materials, we must consider the mass of each part and work with mass-moments.

Again the position of the centre of gravity is given by a similar expression,

$$\bar{x} = \frac{\Sigma(mx)}{\Sigma m}$$

Example 29

Find the centre of gravity for the shaft shown in Diagram 4·54 when part 1 is made from aluminium of density 2·7 g/cm³, and part 2 is made from steel of density 7·7 g/cm³. (Take the sizes on the diagram to be in centimetres.)

Solution

Taking mass-moments about O (remember, mass = volume × density):

Part	Volume (cm³)	Density	Mass (m)	x	mx
1	75·4 (cm³)	2·7	204 g	3	612
2	15·7 (cm³)	7·7	121 g	8·5	1029
Total	–	–	Σm = 325 g	$\bar{x}$	Σmx = 1641

$$\text{Therefore } \bar{x} = \frac{\Sigma mx}{\Sigma m} = \frac{1641}{325} = 5{\cdot}05 \text{ cm}$$

therefore centre of gravity is 5·05 cm from O along OX.

4.9 Stability

The position of the centre of gravity of a body is very important. So is the size of the base of a body which is free-standing. It is the relationship between position of centre of gravity and base size which designers are concerned with when designing bodies for particular purposes. Double-decker buses have a very low centre of gravity and a large wheel-base. The bus can be filled to capacity with passengers standing on the lower deck and it will still be stable. However, passengers are not allowed to stand on the upper deck as well because that would raise the position of the overall centre of gravity and could make the bus tend to topple over on corners, or make it nose-dive in an emergency stop.

Cars with roof-racks should not carry large, heavy objects on the roof-rack for similar reasons, especially if there is only the driver in the car.

There are three kinds of equilibrium:

Stable equilibrium

A body is said to be in a stable equilibrium if it would tend to return to its original position after being tilted slightly. (Wide base and low centre of gravity.)

Unstable equilibrium

A body is said to be in unstable equilibrium if it would tend to topple over after being tilted slightly. (Narrow base and high centre of gravity.)

Neutral equilibrium

A body is said to be in neutral equilibrium if it remains in the new position after being moved. (Height of centre of gravity constant and its position is constantly above the base contact point.)

These three kinds of equilibrium are illustrated in Diagram 4.55. We can see the effect of design on the stability of many common objects, whether they are intended to be used when stationary or when moving.

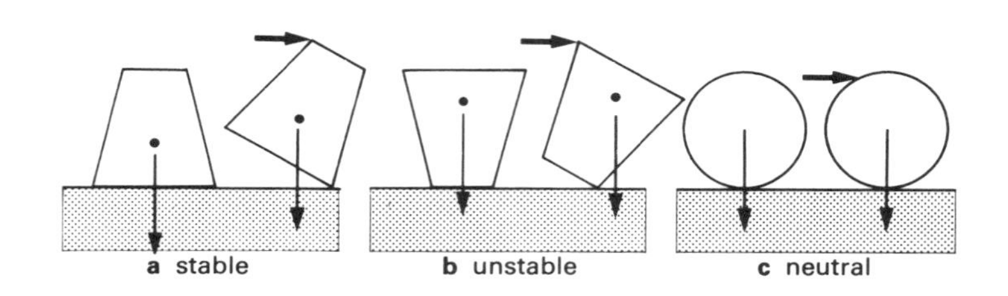

Diagram 4.55

For example look at the special features which give stability to racing cars, land yachts, catamarans, standard lamps, high step-ladders, temporary bus-stop posts.

Also, think about how stability can be altered and improved, e.g. putting ballast in boats, the action of the crew in a sailing dinghy, putting concrete round the base of a clothes-pole. There are many examples of the importance of the position of the centre of gravity.

Consider a man-made reservoir. The water is trapped in a valley by building a dam-wall across the valley exit. The cross-section of such a dam wall may look like that shown in Diagram 4.56**a**.

The diagram shows the direction of the resultant water force and the position of the centre of gravity of the dam wall where its force of gravity acts.

The total force on the dam wall can be found by adding these two forces together on a vector diagram (Diagram 4.56**b**). This gives the resultant force on the wall in magnitude and direction. The exact position in the wall where this resultant acts is the crossing point of the lines of action of the total water force and F_g.

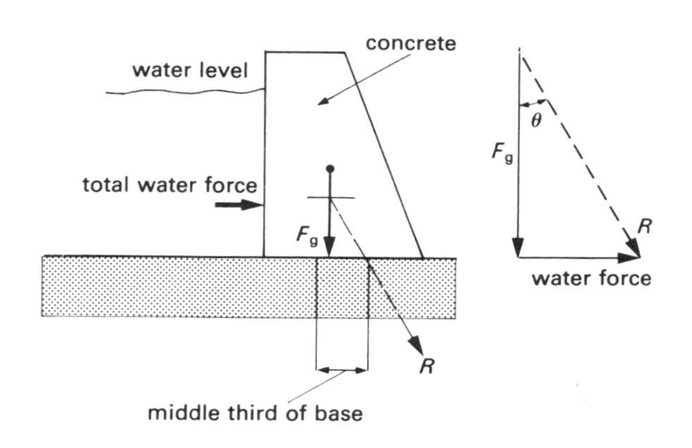

a space diagram (drawn to scale) **b** vector diagram (drawn to scale)

Diagram 4.56

For a concrete dam to be considered stable, the resultant force on the dam must have its line of action passing through the base of the dam. For safety, designers usually make sure that the resultant passes through the 'middle third' of the base.

To work out the forces on a dam wall we will consider 1 metre length of the wall and deal with the forces for that 1 metre length.

Example 30

A gravity masonry dam has a cross-section as shown in Diagram 4.57**a**. It resists a total water force of 2000 kN/m length as shown.

i How far is the centre of gravity of the cross-section from the vertical face of the dam?

ii What weight of dam (per metre length) would just stop the wall overturning?

iii What weight of dam would just keep the resultant force within the middle third of the base?

Solution

i To find the position of the centre of gravity from the vertical face: Consider the cross section to be in two parts, a rectangle and a triangle. Therefore taking area-moments about the vertical face:

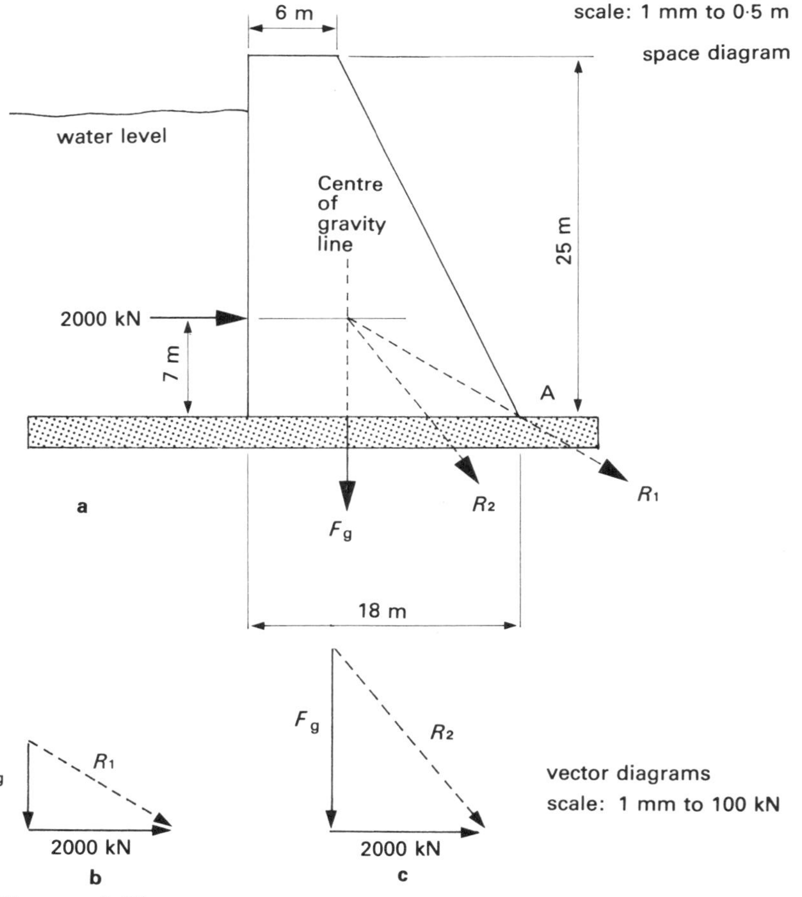

Diagram 4.57

Part	Area (A)	x	Ax
Rectangle	150	3	450
Triangle	150	10	1500
Total	300	$\bar{x}$	1950

Therefore $\bar{x} = \frac{\Sigma(Ax)}{\Sigma A}$

$= \frac{1950}{300}$

$\bar{x} = 6{\cdot}5$ m from the vertical face

ii To find the weight (per metre length) to prevent the dam overturning. For this condition, the resultant force on the wall must be such that its line of action passes through point A. Therefore, on the space diagram (drawn accurately to scale) we can draw the line of action for R_1 as shown in Diagram 4.57**a**. Using the water force as the starting vector, we can draw the vector diagram as in Diagram 4.57**b**.
Therefore from the vector diagram,

$F_g = 1220$ kN/m length

iii Similarly to find the weight (per metre length) to keep the resultant within the middle third of the base, we draw the vector diagram for R_2 as shown in Diagram 4.57**c**.
Therefore, from this vector diagram,

$F_g = 2550$ kN/m length
(Note the similar triangles in space diagram and vector diagram. Therefore this problem could have been done by 'proportionate sides of similar triangles'.)

Suspended bodies

The position of the centre of gravity affects how a body will hang from a single rope or suspension point. The action of gravity is to pull the centre of gravity of a body as low as possible. The lowest possible position is always vertically below the suspension point. Think what happens if you hold a ruler lightly by one end, or a photograph by a corner. The bodies swing until their centres of gravity are vertically below the suspension point. Knowing where the centre of gravity is, lets us know where to hang a body by, to keep it in a particular position.

Exercises

1 Represent each of the following forces by a vector on its own vector diagram drawn to a suitable scale.
 a 20 N acting from West to East.
 b 4 N acting vertically down.
 c 300 kN acting North West.
 d 5 MN acting South East.

2 Find the resultant force for each of the force systems shown in Diagrams 4.58**a** – **d**.

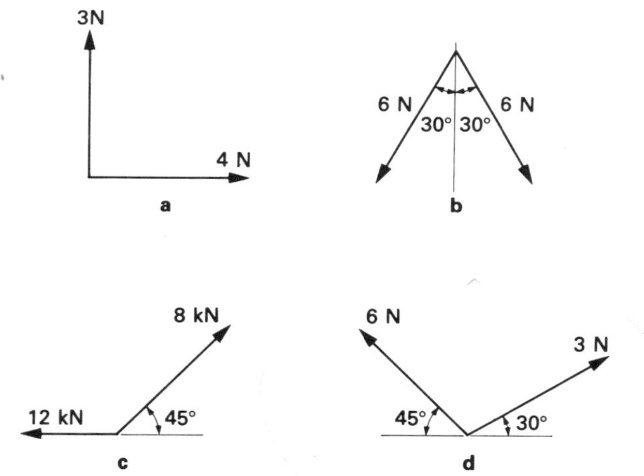

Diagram 4.58

3 Resolve the forces shown in Diagrams 4.59**a** – **d** into their rectangular components horizontal (H) and Vertical (V).

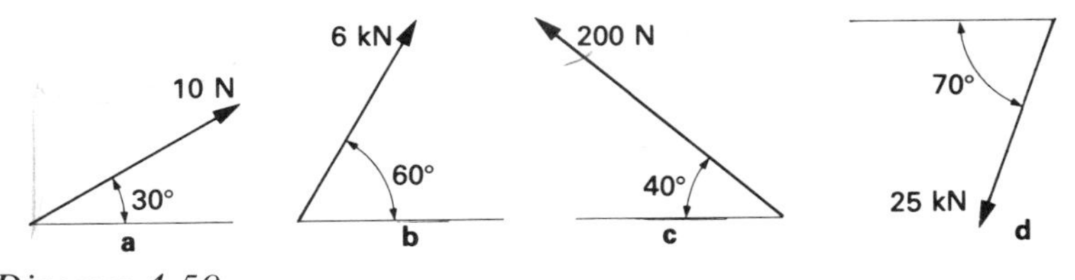

Diagram 4.59

4 Find the equilibrant for each of the force systems shown in Diagrams 4.60**a** – **d**.

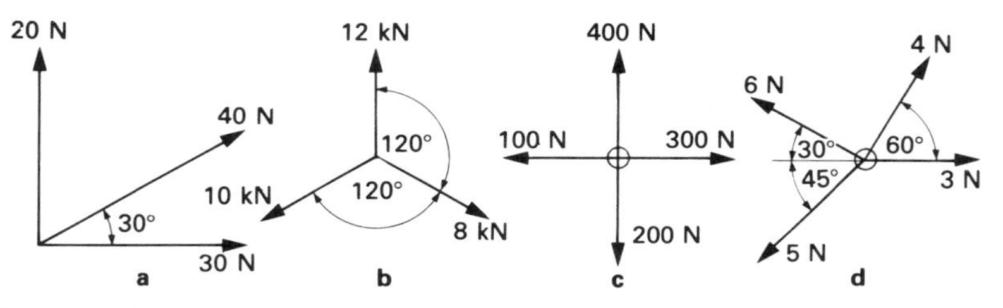

Diagram 4.60

5 For each of the loaded frame structures shown in Diagrams 4.61**a** – **f**, find **i** the magnitude and nature of the forces in each member of the frame, and **ii** the reactions at the supports X and Y.

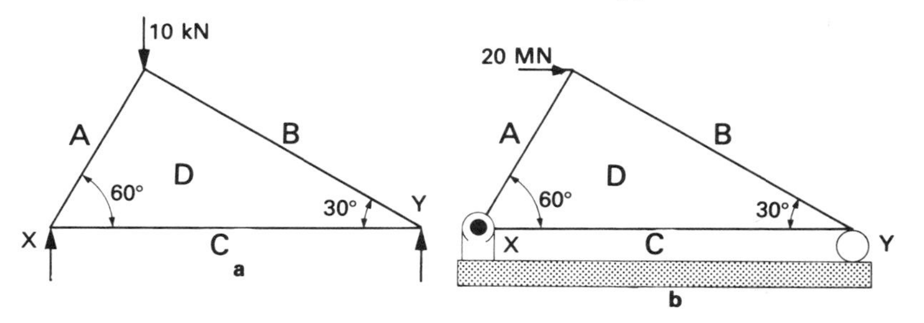

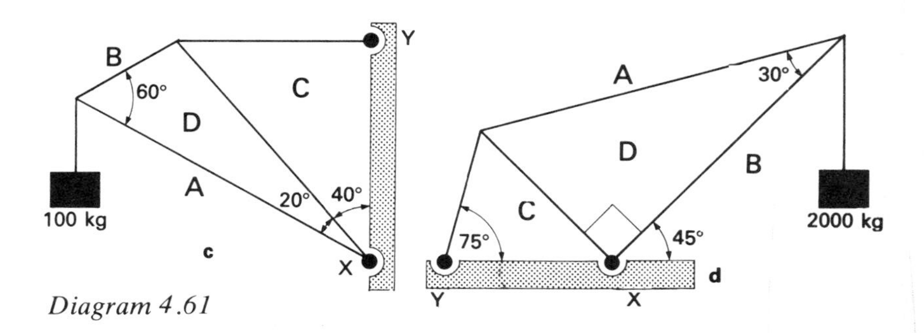

Diagram 4.61

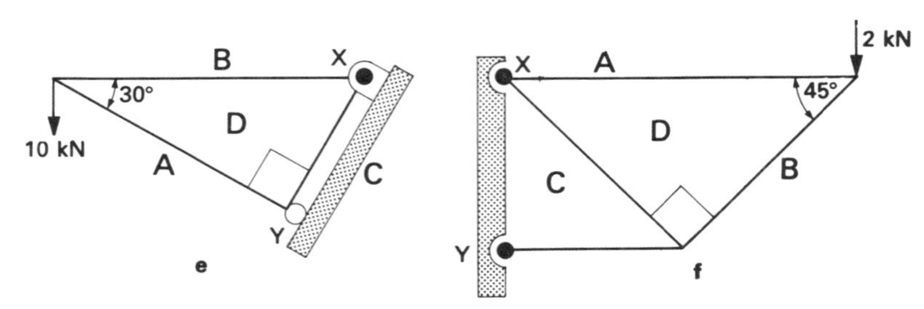

Diagram 4.61 (continued)

6 Find the unknown quantity so that each of the lever systems shown in Diagrams 4.62**a** – **d** is in equilibrium.

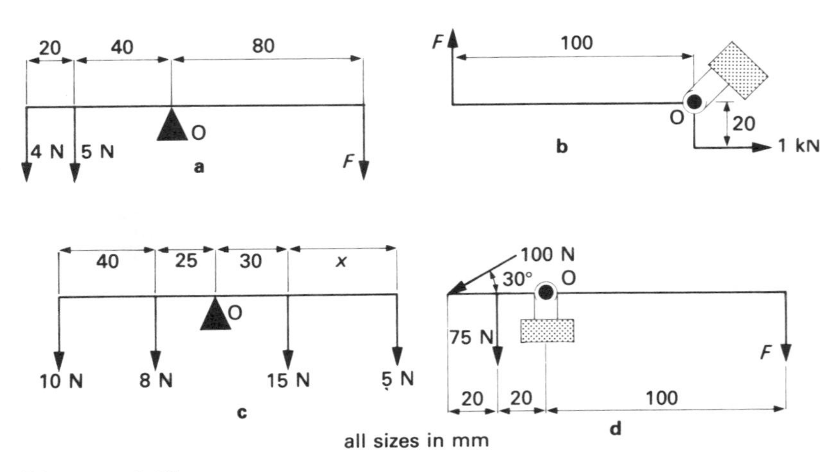

Diagram 4.62

7 Find the reactions at the supports X and Y for each of the loaded beams shown in Diagrams 4.63**a** – **d**.

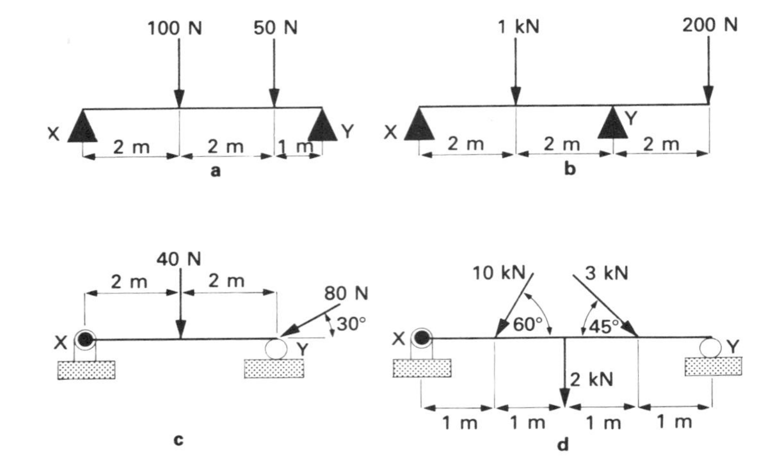

Diagram 4.63

8 A force of 20 N acts on the rim of a pulley which is 80 mm in diameter. This pulley is connected to a 25 mm diameter shaft. Find **i** the torque on the shaft, and **ii** the force which can be transmitted at the surface of the shaft.

9 Find the centroid for each shape shown in Diagrams 4.64**a** – **d**.

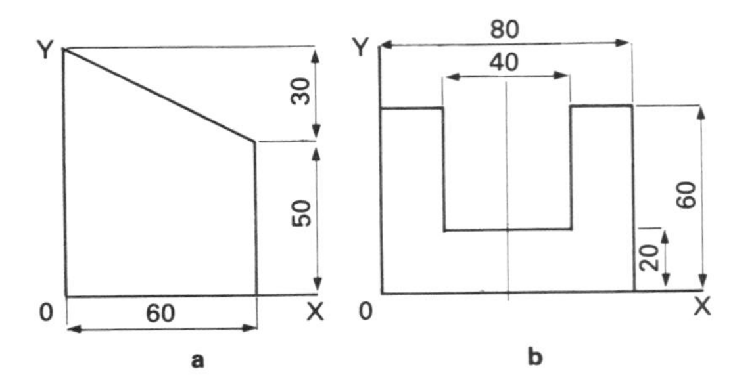

Diagram 4.64

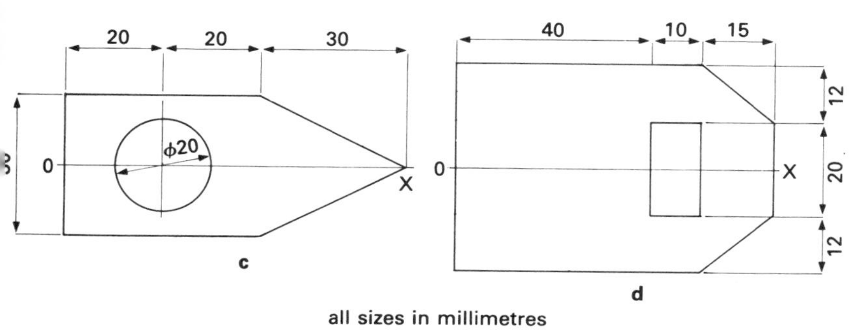

Diagram 4.64 (continued)

10 Find the centre of gravity for each of the solid objects shown in Diagrams 4.65 **a** and **b**. (The two objects are made from the same material.)

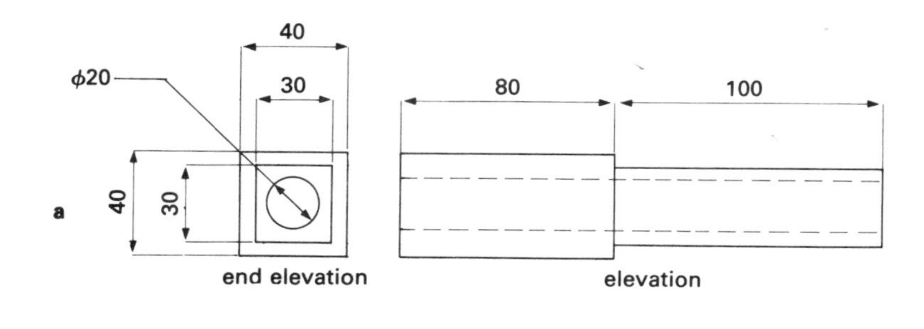

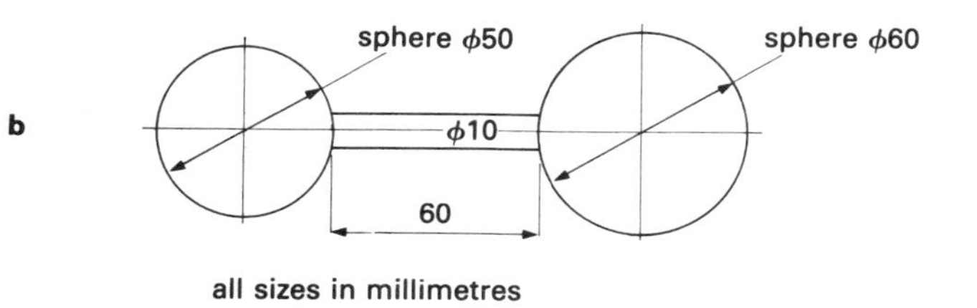

Diagram 4.65

11 The cross section of a masonry dam is shown in Diagram 4.66. The total water force per metre length is 700 kN acting at a height of 4 m up the vertical face of the dam. Find **i** the position of the centroid of the section, and **ii** the weight per metre length of the dam if it is just on the point of overturning.

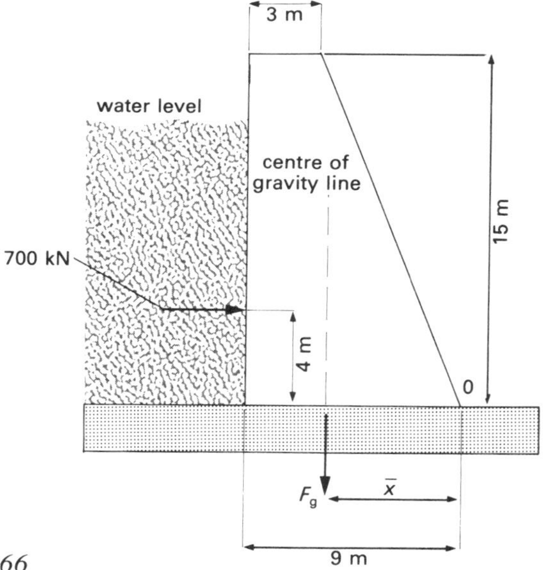

Diagram 4.66

12 The template shown in Diagram 4.64**b** is to be suspended by its top left hand corner. Find the angle which the edge OX will make to the horizontal while the template is being suspended by the corner.

Chapter 5

Dry Friction

5.1 What is friction?

There are many meanings of the word 'friction' today, but in engineering, friction is defined as the resistance of surfaces to sliding over one another.

This can be very useful in everyday life as well as in engineering in particular. If it was not for friction, our shoes would constantly slip when walking, brakes on vehicles would not work and tyres would slip on the roads. It is not difficult to think of many more examples of the usefulness of friction.

Friction can be undesirable too. When pulling a sledge, pushing a barrow or opening a door we must use sufficient force to overcome the frictional resistance to motion before the body will move.

A six-car accident during the 1980 Long Beach Grand Prix

5.2 Normal reaction and force of friction

Consider a body of mass m resting on a horizontal surface which is not perfectly smooth (Diagram 5.1).

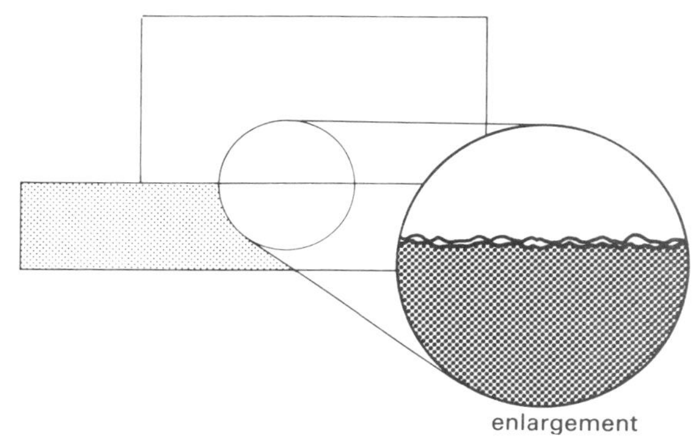

Diagram 5.1

Due to gravity, F_g acts vertically down and it can be considered to act at the centre of gravity of the body. Also, because the body is resting on a surface, the surface reacts to the gravitational force with an equal but opposite force to maintain equilibrium. This reaction (symbol R) can also be considered to act in a straight line through the centre of gravity of the body as shown in Diagram 5.2. This is usually drawn as in Diagram 5.2**b**.

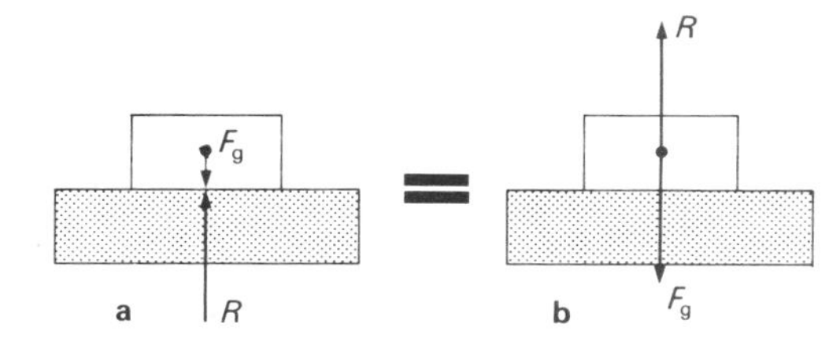

Diagram 5.2

Let a small horizontal force, F, be applied to the body, but not big enough to make the body move. The body now has three forces acting on it, F, F_g and R. These three forces must be in equilibrium because the body is at rest. Since the forces are not parallel, they must be concurrent and also their vectors in the vector diagram must form a closed triangle. We know the directions and magnitudes of F and F_g, therefore it is the surface reaction R which must change to take up a position which satisfies the conditions of equilibrium.

The only possible arrangement is shown in Diagram 5.3 along with the appropriate vector diagram.

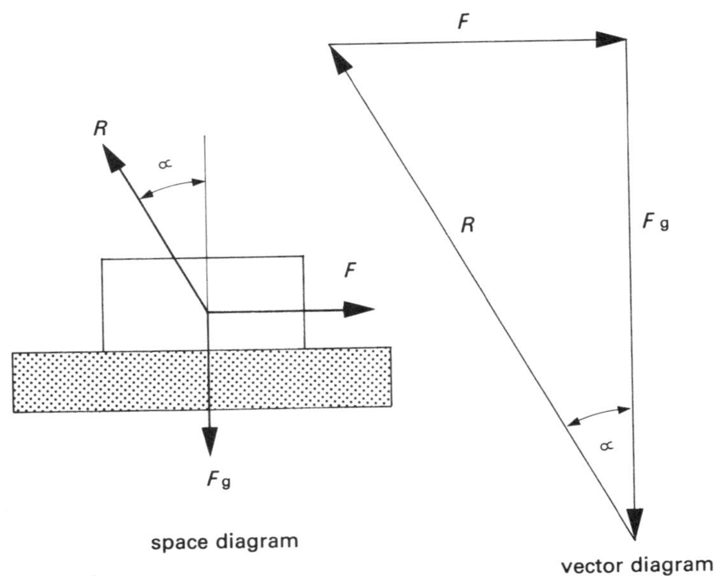

Diagram 5.3

From the space diagram we can see that R now makes an angle α to the vertical, in order to give a closed triangle for the vector diagram.

This must be true because the surface reacts to a vertical force F_g **and** a horizontal force F at the same time. Therefore R can be considered to be made up of two components, **i** a component at right angles to the surfaces in contact, to react to F_g, and **ii** a component parallel to the surfaces in contact, to react to the applied force F. The component of R at right angles to the surfaces in contact has not changed because nothing has changed the effect of F_g, but the component of R parallel to the surfaces in contact is due entirely to the applied force F. This component parallel to the surfaces in contact is the resistance of the surfaces to sliding over one another and **always** opposes the direction of motion (or intended motion). Therefore, the overall surface reaction R will always slope in the opposite direction to motion or intended motion.

The two components of R are given particular names:

i The component of R at right angles to the surfaces in contact is called the **normal reaction** and is given the symbol R_n. (Normal means at right angles.)

ii The component of R parallel to the surfaces in contact is called the **force of friction** and is given the symbol F_f.

Diagram 5.4 shows how the overall reaction R can be replaced by its components F_f and R_n. We must remember that R_n and F_f are components of one force, R, and they are not separate forces in their own right.

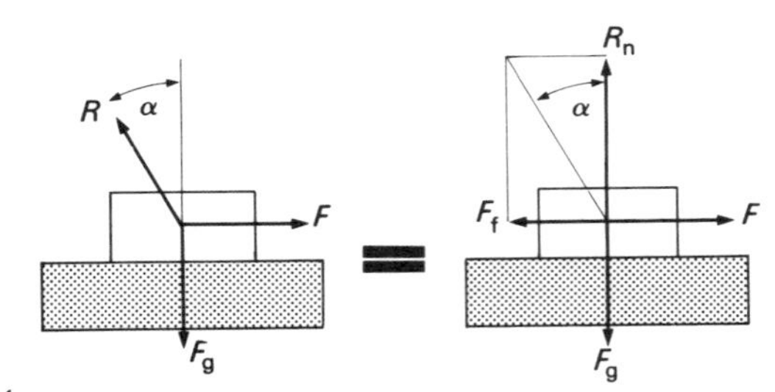

Diagram 5.4

Also note that F_f opposes the direction of intended motion due to the applied force.

Now let the small horizontal force, F, increase slowly until the body is just on the point of moving. Again nothing has happened to change F_g therefore the normal component of R must stay the same (R_n). But the parallel component F_f must have increased to balance the effect of the increasing applied force F. Therefore the overall surface reaction R

must be greater and at a larger angle to the perpendicular to the surface (Diagram 5.5).

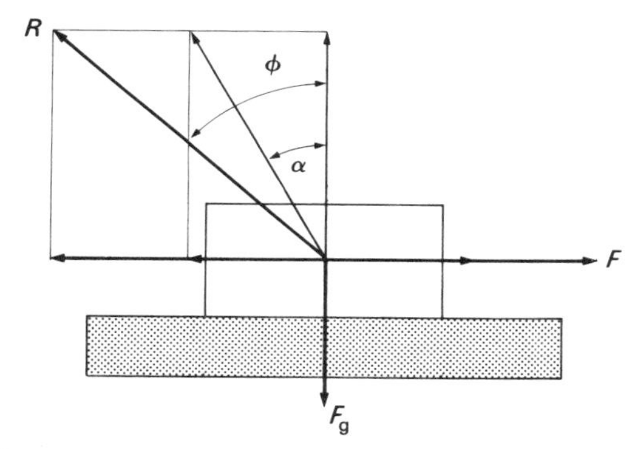

Diagram 5.5

Because the body is on the point of moving the surfaces in contact have reached their limit of resistance to motion. If the applied force is increased again then the body will move. This critical point is very important. We have seen how the friction force component of R has increased from 0 up to its maximum or limiting value. This means that the overall reaction R has increased to its limit too, and that the angle which R makes with the perpendicular to the surface is also at a limit. In this situation where the body is still static (i.e. at rest) but only just, the force of friction is called the **limiting value of static friction.** Therefore the **limiting angle of static friction** is the largest angle which can exist between the overall surface reaction R and the perpendicular to the surface, and this occurs when the body is just on the point of moving along the surface. The limiting angle of static friction is given the symbol ϕ (Greek letter phi) or sometimes ϕ_s for reasons which will be made clear later in the chapter.

Diagram 5.6 shows R and its angle increasing to their limiting values.

By experiment we could show that once a body has started to move it can be kept moving by a force which is slightly less than the force needed to make it move in the first place. This means that the resistance to motion must be less for bodies which are sliding than for bodies which are at rest. The word 'kinetic' means 'of motion' therefore we can say that the limiting value of static friction is always greater than the kinetic friction.

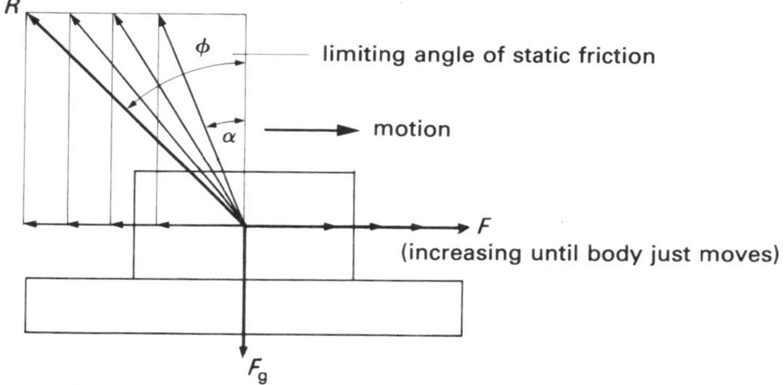

Diagram 5.6

For a body which has started moving the friction force component of R is reduced. Therefore R and its angle must also be reduced. This reduced angle is called the angle of kinetic friction and is given the symbol ϕ or sometimes ϕ_k to distinguish it from the limiting angle of static friction ϕ_s. Therefore ϕ_s is greater than ϕ_k.

For the same body on a different kind of surface, different values of resistance to motion will be found. Therefore friction is dependent on the nature of surfaces in contact, e.g. a body on a rough wooden surface would experience greater resistance to motion than on a polished glass surface.

For different surfaces in contact, there will be different values of friction. These values are normally given in terms of the limiting angle of static friction as found by experiment for particular surface combinations, e.g.

i ϕ_s for steel on graphite $= 5°$
ii ϕ_s for wood on metal $= 18°$
iii ϕ_s for steel on steel $= 12°$

5.3 Coefficient of friction

Consider the two components of the surface reaction R, i.e. normal reaction R_n and friction force F_f. The triangle shown in Diagram 5.7 is a right-angled triangle. Therefore, from mathematics

$$\tan \phi = \frac{F_f}{R_n}$$

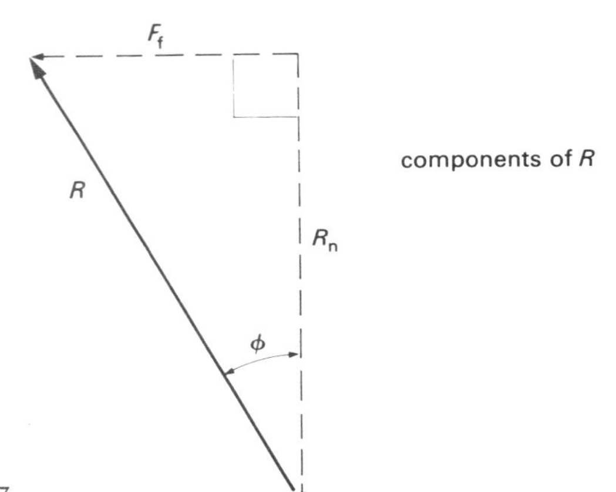

Diagram 5.7

This relationship will always apply and therefore if we know the angle of friction ϕ then we will know the shape of the right-angled triangle which gives F_f and R_n as well as R.

The quantity $\tan \phi$ is called the **coefficient of friction** which is given the symbol μ (Greek letter mu).

In general $\tan \phi = \mu$
$\tan \phi_s = \mu_s$, the coefficient of limiting static friction,
and $\tan \phi_k = \mu_k$, the coefficient of kinetic friction.

The coefficients of friction, μ, for our purposes will always be a decimal fraction smaller than 1, e.g.

i ϕ_s for steel on graphite $= 5°$
therefore $\tan \phi_s = \tan 5° = 0{\cdot}09$
therefore μ_s for steel on graphite $= 0{\cdot}09$

ii ϕ_s for wood on metal $= 18°$
therefore $\tan \phi_s = \tan 18° = 0{\cdot}33$
therefore μ_s for wood on metal $= 0{\cdot}33$

iii ϕ_s for steel on steel $= 12°$
therefore $\tan \phi_s = \tan 12° = 0{\cdot}213$
therefore μ_s for steel on steel $= 0{\cdot}213$

However, $\tan \phi = \dfrac{F_f}{R_n}$, from Diagram 5.7
and $\mu = \tan \phi$
therefore in general, $\mu = \dfrac{F_f}{R_n}$

or in words, $\text{coefficient of friction} = \dfrac{\text{friction force}}{\text{normal reaction}}$

This is usually written as $F_f = \mu R_n$

In this form it is easy to see that the frictional resistance to motion depends on two things: **i** the nature of the surfaces in contact (which determines μ) and **ii** the normal reaction between the surfaces in contact (R_n). Also the size of the friction force is directly proportional to the normal reaction ($F_f \propto R_n$), i.e. if the normal reaction is doubled then the force of friction will be doubled too.

5.4 Summary

We must remember certain things about friction:

1. Friction opposes motion.
2. The overall reaction R has two components F_f and R_n.
3. ϕ_s is the limiting angle of static friction.
4. ϕ_k is the angle of kinetic friction.
5. In general $\mu = \tan \phi$.
6. $F_f = \mu R_n$.

Example 1

A wooden crate has a mass of 15 kg and is to be pulled along a wooden floor by a horizontal force. The limiting angle of static friction for wood on wood is 22°. Find the force which will start the crate moving.

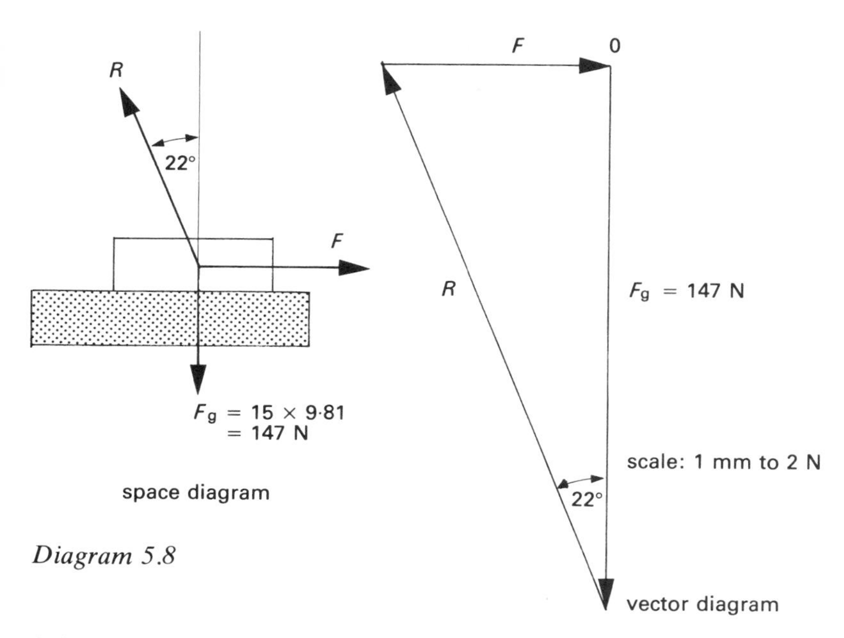

Diagram 5.8

Solution

The first step in any problem is to draw a diagram, and put the information from the question on the diagram. The space diagram in Diagram 5.8 shows all the information from the question:

i magnitude and direction of F_g

ii direction of F

iii direction of R

Although the magnitudes of F and R are not known, there is enough information to draw the vector diagram as shown in Diagram 5.8 as follows:

i from a suitable starting point O, draw the vector for F_g vertically down to a suitable scale;

ii from the lower end of vector F_g draw a line at 22° to F_g for the direction of R;

iii from the upper end of vector F_g draw the horizontal line for the direction of F.

We can now see that we have a closed triangle. The value of F can be found by measuring the length of the horizontal vector (to scale), i.e. 30 mm representing a force of 60 N. Therefore, the force necessary is 60 N$\rightarrow$. (Note that 60 N must be the limiting value of static friction, F_f, in this case.)

Example 2

Find the horizontal force which will make a mass of 4 kg move along a horizontal surface. The coefficient of limiting static friction, μ_s, is 0·6 for the surface in contact.

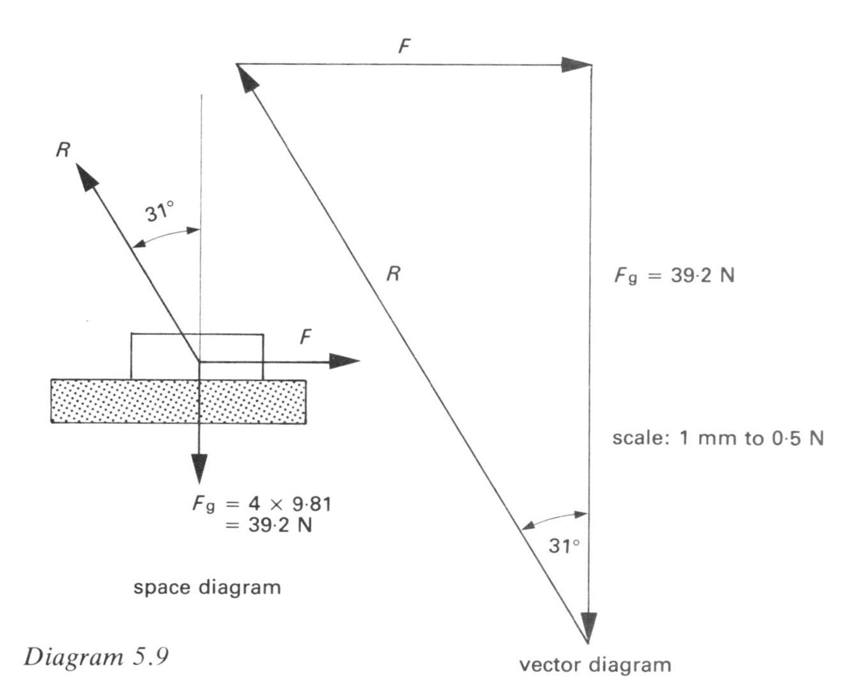

Diagram 5.9

Solution

Again take all the information from the question and draw a sketch. In this case we are given a coefficient of friction and we need ϕ_s.

Therefore, remembering that $\mu = \tan\phi$
then $\tan\phi_s = \mu_s = 0{\cdot}6$

By looking up the table of values for Natural Tangents in a book of mathematical tables, we would find that the value closest to 0·6 is for an angle of 31°. Therefore $\phi_s = 31°$.

We can now draw the diagrams as shown in Diagram 5.9. From the space diagram, the vector diagram can be drawn following the same method as in Example 1; i.e. draw F_g first, add on a line for the direction of R at 31° and finally draw the horizontal vector for F. From the vector diagram, the length of $F = 48$ mm and from the scale of 1 mm to 0·5 N, then $F = 24$ N. Therefore force necessary is 24 N →. (Again this is equal but opposite to F_f.)

5.5 Inclined forces

So far we have only considered bodies which have been made to move by horizontal forces. The study of inclined forces follows a very similar pattern.

Consider a body of mass m resting on a horizontal surface subjected to an applied force F which makes an angle α to the horizontal as shown in Diagram 5.10. Let the body be just on the point of moving, and let the limiting angle of static friction be ϕ_s.

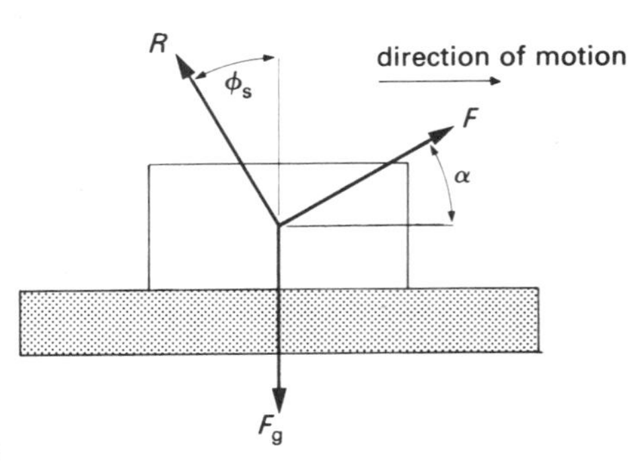

Diagram 5.10

The way of solving this problem is exactly the same as in the previous problems except that F is inclined instead of horizontal. The vector diagram would be as shown in Diagram 5.11.

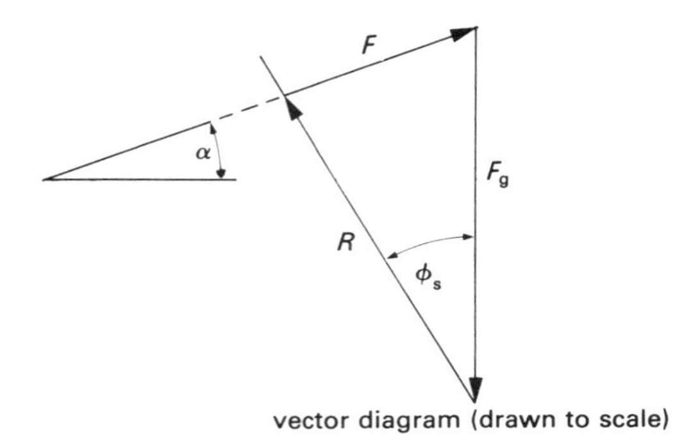

Diagram 5.11

Note that because F is inclined, it is no longer equal and opposite to F_f; nor is R_n equal and opposite to F_g. If we consider the applied force, F, to be made up of two components we can see why. F has a horizontal component and a vertical component as shown in Diagram 5.12.

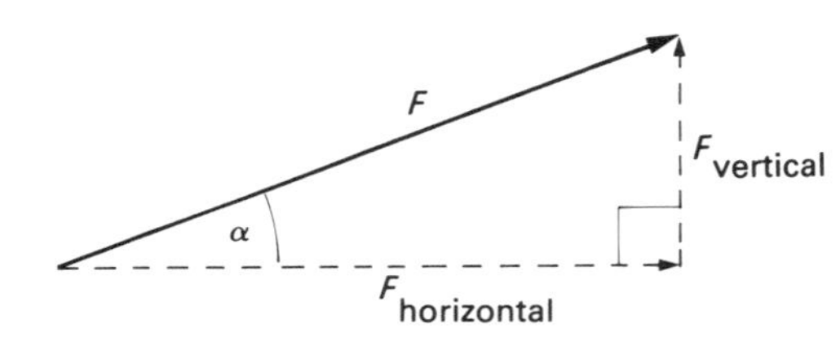

Diagram 5.12

The horizontal component of F is the one which is equal and opposite to the horizontal component of R. Therefore the friction force is overcome by the horizontal component of the applied force. The vertical component of F acts opposite in direction to F_g and therefore it lightens the load on the surface. Hence the surface has to react to a lighter load.

Therefore $R_n = F_g -$ vertical component of F.

Putting the components of F and R on a diagram gives Diagram 5.13.

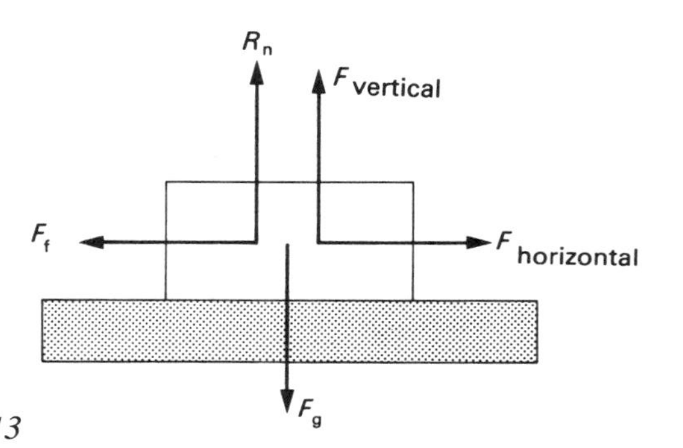

Diagram 5.13

Example 3

A loaded sledge has a mass of 60 kg. It is pulled along the ground by a rope which makes an angle of 30° to the ground. The coefficient of kinetic friction, μ_k, is 0·25 for the surfaces in contact. Find the tension in the rope.

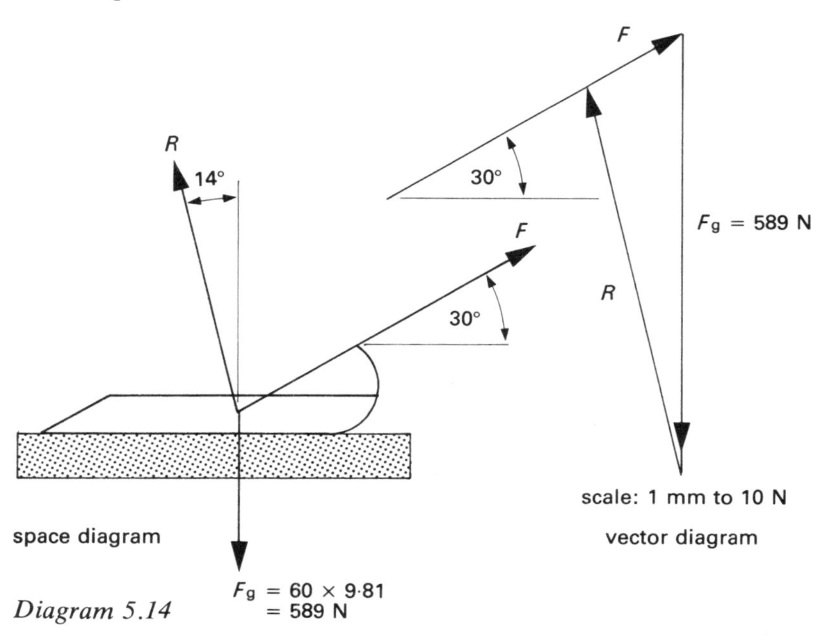

Diagram 5.14

Solution

Let the tension in the rope be F.

$$\tan \phi_k = \mu_k = 0{\cdot}25$$

therefore ϕ_k = 14° from tables.

Diagram 5.14 shows the space diagram with the information from the question. From the space diagram the vector diagram is drawn starting with F_g, then the direction of R at 14° to F_g, and finally the direction of F at 30° to the horizontal.

As before, the scaled length of F gives its magnitude. Therefore from the vector diagram the scale length of F is 15 mm, therefore F = 150 N, i.e. tension in the rope is 150 N.

Example 4

A man pushes a box of mass 8 kg along a level path. His arms make an angle of 15° to the horizontal and the angle of kinetic friction, ϕ_k, is 20° for the surfaces in contact. Find the force in the man's arms.

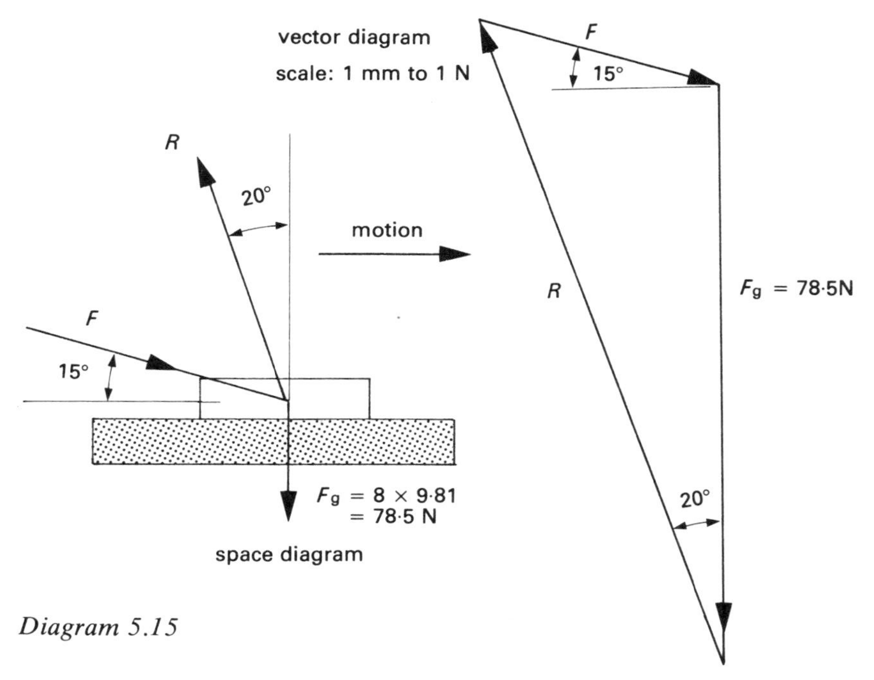

Diagram 5.15

Solution

From the vector diagram in Diagram 5.15 the scaled length of F is 33 mm, i.e. F = 33 N. Therefore force in man's arms is 33 N.

5.6 Inclined planes

Consider a body of mass m resting on a surface which is inclined at an angle θ to the horizontal; the angle θ being small enough so that the body does not slide down the surface due to gravity.

Since the body is at rest under the action of gravity then the other force, which is the surface reaction, must be maintaining equilibrium, i.e. R must be equal and opposite to F_g as shown in Diagram 5.16. The body does not slide down the surface because the frictional resistance between the surfaces is big enough to prevent motion.

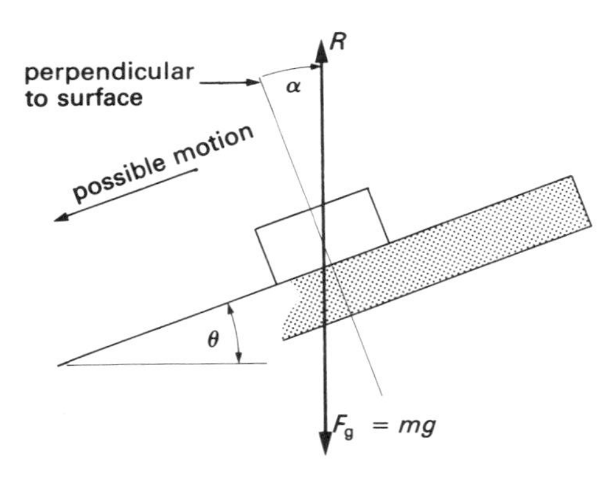

Diagram 5.16

Diagram 5.16 shows R directly opposite in direction to F_g. It also shows the angle which R makes to the perpendicular to the surface (α). Since R has been able to take up this position, then the angle α must be less than the limiting angle of static friction ϕ_s. (Remember that R can make any angle with the perpendicular to a surface up to the limiting angle of static friction. This applies to horizontal and inclined surfaces.)

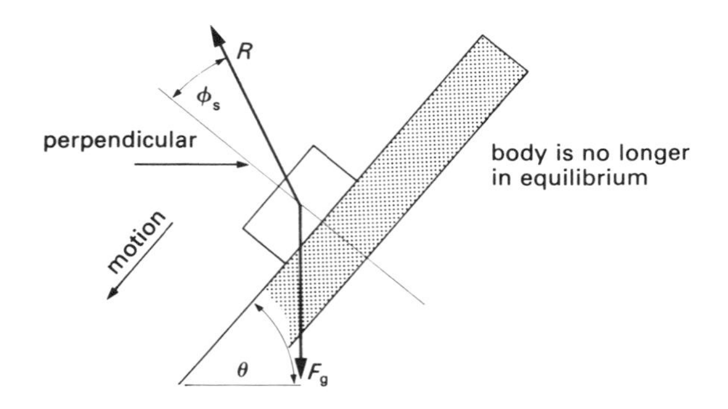

Diagram 5.17

There must be a point where the steepness of the slope is so great that R cannot swing far enough to balance F_g, as in Diagram 5.17.

We can see that with R at its limiting angle of static friction ϕ_s, it cannot balance the effect of gravity F_g. Therefore the body will slide down the surface due to gravity.

The critical point arrives when the angle of the surface to the horizontal is exactly equal to the limiting angle of static friction. (This angle of tilt is sometimes called the **angle of repose**, but this name is more correctly used for the steepest possible slope made by a pile of granular material like sand or gravel.)

Therefore, when $\theta = \phi_s$ then the body is on the point of sliding down the surface due to gravity.

Example 5

A post-office parcel chute is to be made of steel and inclined at such an angle that the parcels will just slide down by themselves due to gravity. The coefficient of kinetic friction, μ_k, for the surfaces in contact is 0·7. What is the necessary angle for the chute?

Solution

$$\tan \phi_k = \mu_k = 0{\cdot}7$$

therefore $\phi_k = 35°$ from tables

Also at the critical point $\theta = \phi_k$

therefore chute must be at least 35° to the horizontal.

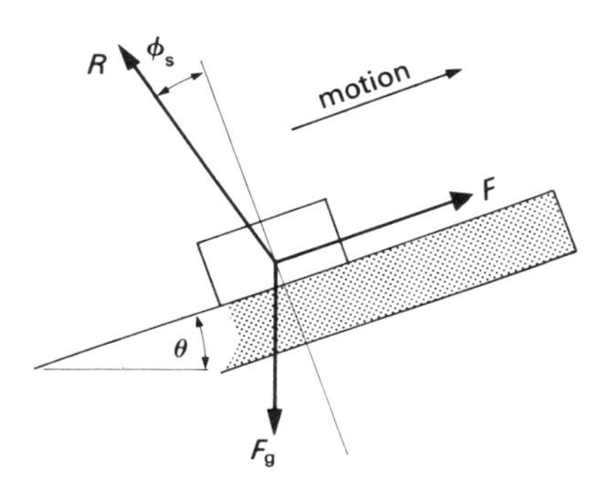

Diagram 5.18

Consider a body of mass m resting on an inclined surface as shown in Diagram 5.18. Let a force F, parallel to the surface, be applied to the body so that the body is just on the point of moving up the slope.

Since the surface reacts to applied forces, R must react in such a way that it opposes the direction of motion (see Diagram 5.18). To find the magnitude of F, we do the same as before, i.e. draw the vector diagram for the three force system (F, F_g and R) since we know all the directions and the magnitude of F_g. This vector diagram is shown in Diagram 5.19.

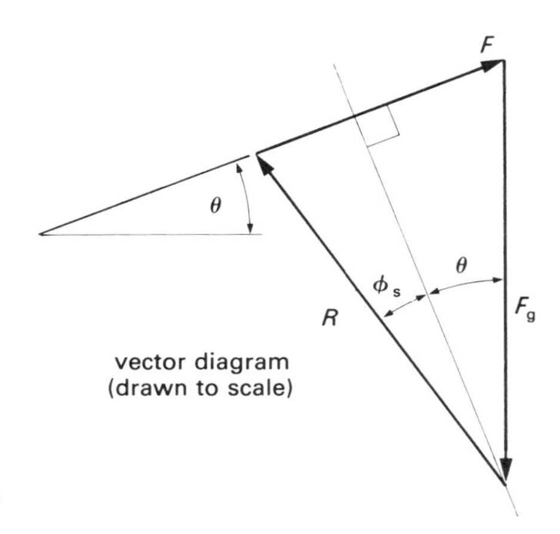

Diagram 5.19

Example 6

A concrete slab of mass 50 kg is being pulled up a driveway by a rope parallel to the slope. The angle of the driveway is 20° and the coefficient of kinetic friction for the surfaces in contact is $\mu_k = 0{\cdot}6$. Find the force in the rope.

Solution

From the vector diagram in Diagram 5.20 the scaled length of F is 44 mm. Therefore $F = 440$ N, i.e. force in rope is 440 N.

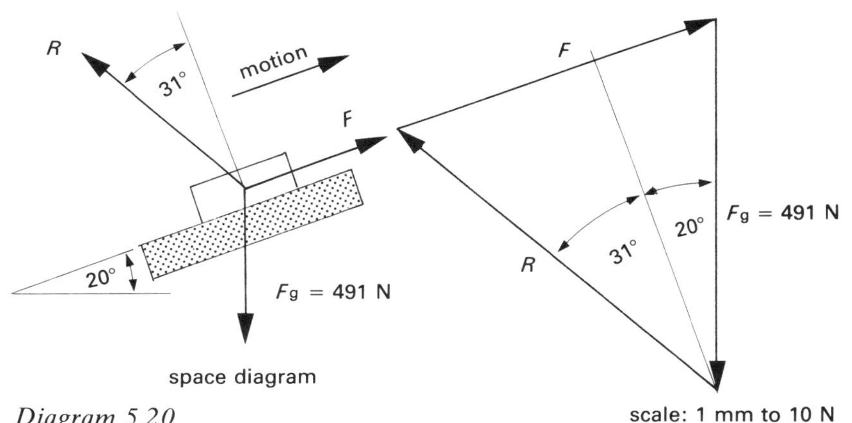

Diagram 5.20

Example 7

Find the force to push a sledge of mass 20 kg down a slope of 10° when the angle of kinetic friction for the surfaces in contact is $\phi_k = 22°$. (Assume that the force is parallel to the slope.)

Solution

The method is exactly the same. We must be careful when drawing the space diagram to draw R opposing the direction of motion. In this example the motion is down the plane, therefore R must be inclined up the plane. (The word 'plane' is very often used instead of the word 'surface'.)

The vector diagram in Diagram 5.21 shows that vector R makes an angle of 12° to vector F_g. To get this position for R in the vector diagram, the position of the perpendicular to the surface was drawn

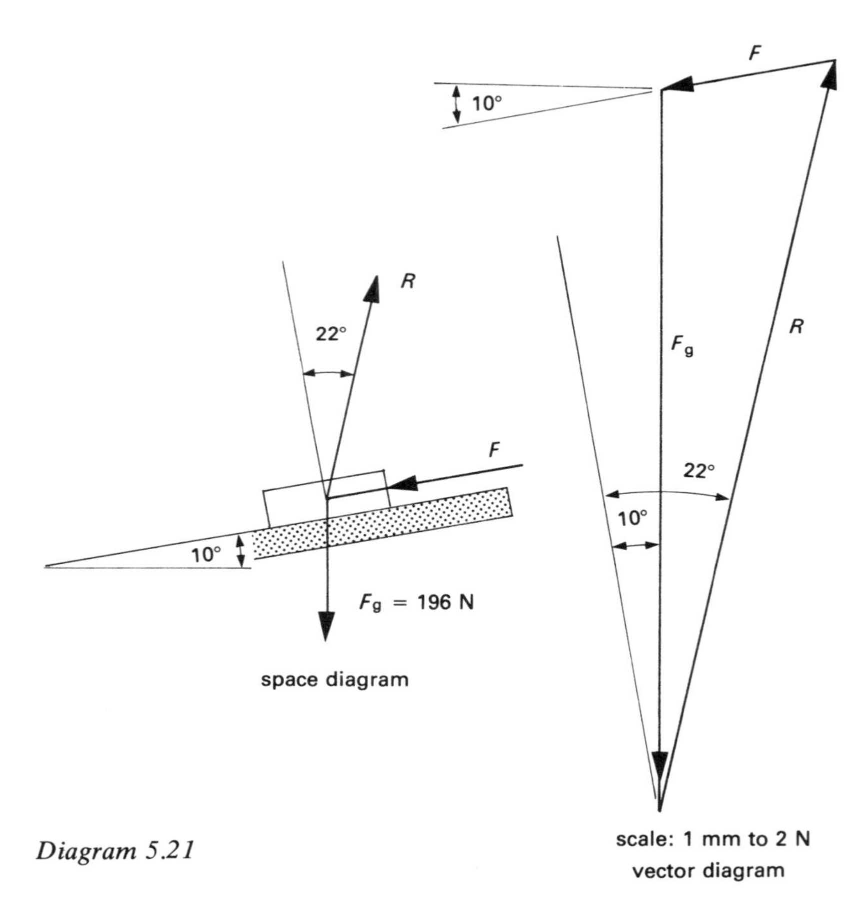

Diagram 5.21

first, at 10° to vector F_g. Then the angle of 22° was measured back to the right. Hence the position of R.

It is worth noting that the angle between vector F_g and the perpendicular to the surface is always exactly the same as the angle that the sloping surface makes to the horizontal.

Returning to the example; from the vector diagram, the scaled length of F is 22·5 mm. Therefore $F = 45$ N, i.e. necessary force down the plane is 45 N.

Example 8

An emergency stretcher is to be lowered by rope down a steep bank. The angle of the bank is 50° and the coefficient of kinetic friction, μ_k, is 0·4. The total combined mass of stretcher and patient is 75 kg. Find the controlling force on the rope so that the stretcher slides gently down the bank.

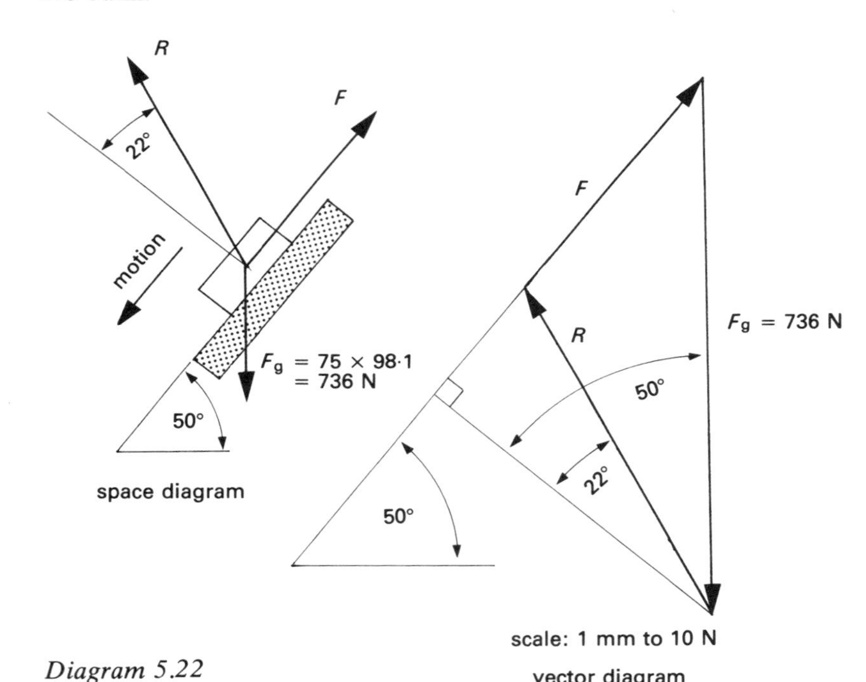

Diagram 5.22

Solution

From the vector diagram, scaled length of F is 37 mm. Therefore $F =$ 370 N, i.e. controlling force on rope is 370 N.

In all these examples the applied force was the unknown. Obviously the same methods can be used when the applied force is known but either the angle of friction (or coefficient of friction) or mass which is sliding is the unknown quantity which has to be found.

Example 9

Find the angle of kinetic friction if it takes a force of 100 N to pull a body of mass 15 kg up a slope of 30°. (The applied force is parallel to the slope.)

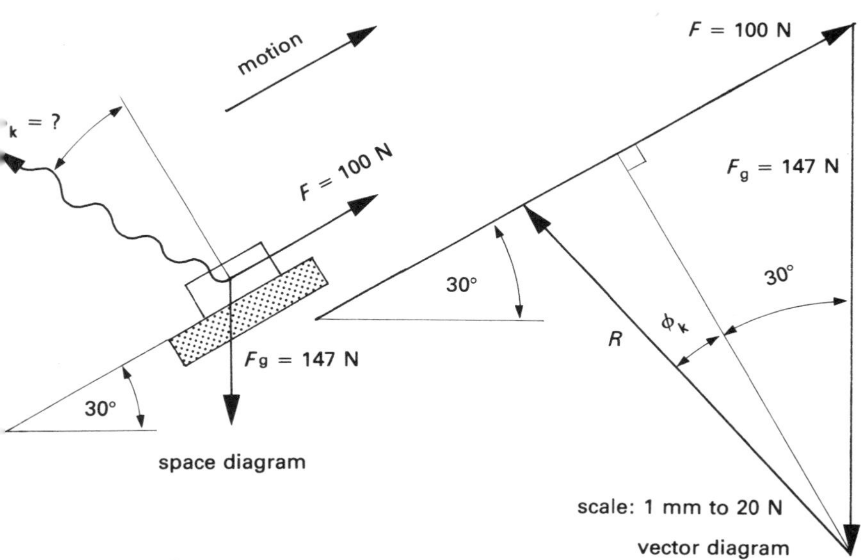

Diagram 5.23

Solution

In this case both F and F_g are known in magnitude and direction. Therefore their vectors can be drawn. Since R is the only remaining force, it must close the vector diagram. Hence ϕ_k can be measured by protractor.

From the vector diagram in Diagram 5.23 the angle between vectors F_g and R is 42°, but this angle is made up of $30° + \phi_k$:

$$30° + \phi_k = 42°$$

i.e. angle of kinetic friction $\phi_k = 12°$

Note that since $\phi_k = 12°$, then $\mu_k = \tan\phi_k = \tan 12 = 0{\cdot}21$.
therefore $\mu_k = 0{\cdot}21$

Example 10

The maximum force which a winch can exert is 600 N. The winch is used to drag crates up a ramp. The coefficient of limiting static friction is 0·23. What is the largest mass the winch can cope with on a slope of 35°?

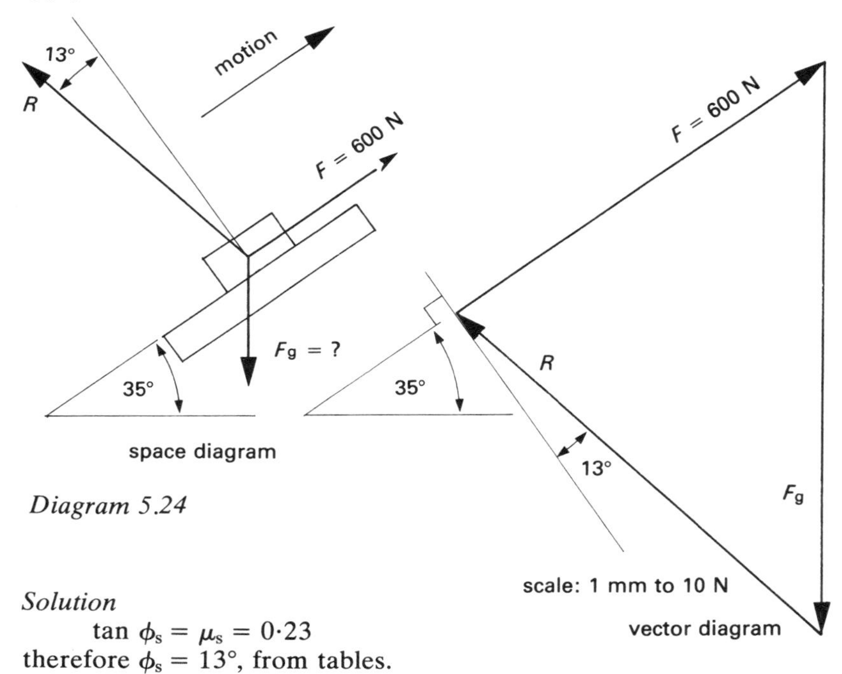

Diagram 5.24

Solution

$\tan\phi_s = \mu_s = 0{\cdot}23$
therefore $\phi_s = 13°$, from tables.

Therefore we now know the magnitude and direction of F, and the directions only for R and F_g. We still have sufficient information to draw the vector diagram as shown in Diagram 5.24. From the vector diagram, the scaled length of F_g is 78 mm.

Therefore $F_g = 780$ N
but, $F_g = m \times g$
therefore $m = \frac{780}{9{\cdot}81}$ kg
and $m = 79{\cdot}4$ kg
Largest mass is 79·4 kg

5.7 Analytical treatment

Instead of dealing with the real forces F, F_g and R in problems on friction, we can consider only those forces which are either parallel or perpendicular to the surfaces in contact. Any other force is divided into its two rectangular components which will be parallel and perpendicular to the surfaces in contact.

For horizontal surfaces

Since the body will either be at rest or moving with uniform motion, the conditions of equilibrium must apply:

i the forces perpendicular to the plane must balance, and
ii the forces parallel to the plane must balance.

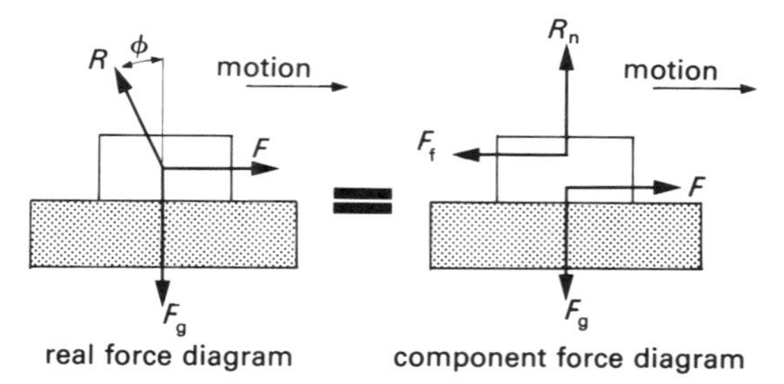

Diagram 5.25

In Diagram 5.25 the surface reaction R has been replaced by its components F_f and R_n. This will always be the case since R always makes an angle with the perpendicular.

Now applying the conditions of equilibrium:

i $R_n = F_g$ and **ii** $F_f = F$

These two expressions have a relationship because, as we have already seen, $F_f = \mu R_n$ (see Section 5.3).

Example 11

A body of mass 10 kg rests on a horizontal surface. The coefficient of limiting static friction, μ_s, is 0·3. Find the horizontal force necessary to make the body move.

Solution

See Diagram 5.26.

For equilibrium, $R_n = F_g$
therefore $R_n = mg$
$= 10 \times 9{\cdot}81$
$= 98{\cdot}1$ N
but $F_f = \mu_s R_n$
therefore $F_f = 0{\cdot}3 \times 98{\cdot}1$
$= 29{\cdot}4$ N
Also for equilibrium, $F = F_f$
$= 29{\cdot}4$ N
therefore necessary horizontal force $= 29{\cdot}4$ N

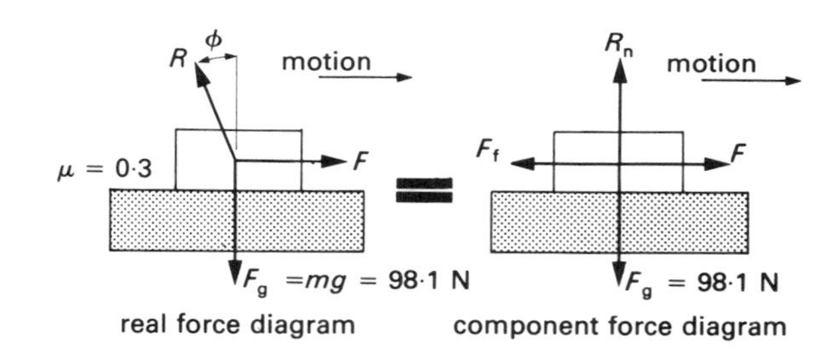

Diagram 5.26

For inclined forces

The component force diagram shows that F has been replaced by its components $F \sin\alpha$ and $F \cos\alpha$ (see Diagram 5.27).

Again, for equilibrium, $R_n + F \sin\alpha = F_g$
and $F_f = F \cos\alpha$
Also remember that $F_f = \mu R_n$

Example 12

A mass of 40 kg is pulled along the ground by a rope which makes an angle of 30° to the ground. The coefficient of kinetic friction, μ_k, is 0·25. Find the force in the rope.

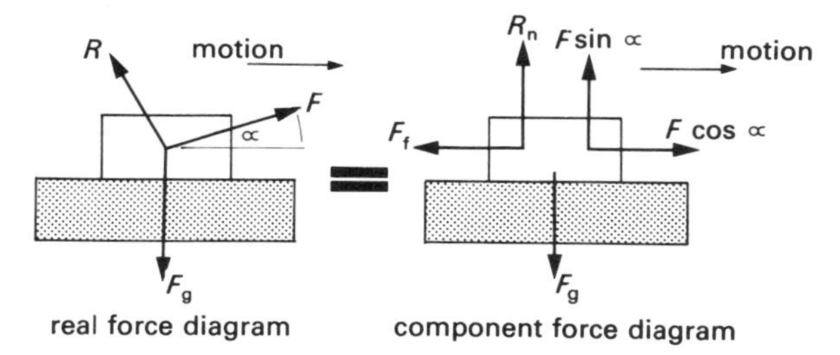

Diagram 5.27

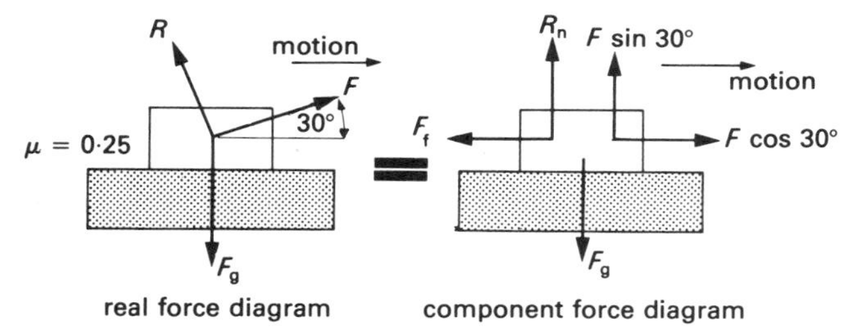

Diagram 5.28

Solution

See Diagram 5.28.

For equilibrium, $R_n + F \sin 30 = F_g$

or

$$\begin{aligned} R_n &= F_g - F \sin 30 \\ &= (40 \times 9{\cdot}81) - F \sin 30 \\ &= 392{\cdot}4 - F \sin 30 \end{aligned}$$

but $F_f = \mu_k \times R_n$

therefore

$$\begin{aligned} F_f &= \mu_k \times (392{\cdot}4 - F \sin 30) \\ &= (0{\cdot}25 \times 392{\cdot}4) - 0{\cdot}25\, F \sin 30 \\ &= 98{\cdot}1 - 0{\cdot}25\, F \sin 30 \end{aligned}$$

Also for equilibrium, $F_f = F \cos 30$

therefore $F \cos 30 = 98{\cdot}1 - 0{\cdot}25\, F \sin 30$

Now, collecting all the terms containing F to one side:

$$F \cos 30 + 0{\cdot}25\, F \sin 30 = 98{\cdot}1$$

Taking F as a common factor gives:

$$\begin{aligned} F[\cos 30 + 0{\cdot}25 \sin 30] &= 98{\cdot}1 \\ F[0{\cdot}866 + (0{\cdot}25 \times 0{\cdot}5)] &= 98{\cdot}1 \\ F(0{\cdot}866 + 0{\cdot}125) &= 98{\cdot}1 \\ F \times 0{\cdot}991 &= 98{\cdot}1 \\ F &= \frac{98{\cdot}1}{0{\cdot}991} \\ &= 99{\cdot}0 \text{ N} \end{aligned}$$

therefore force in the rope is 99·0 N

For inclined planes

On inclined planes, as in Diagram 5.29, the applied forces will always be considered to be parallel to the plane. Therefore the forces which must be divided into their components are R and F_g.

Remember that the equilibrium conditions are applied to the forces parallel to the plane and perpendicular to the plane.

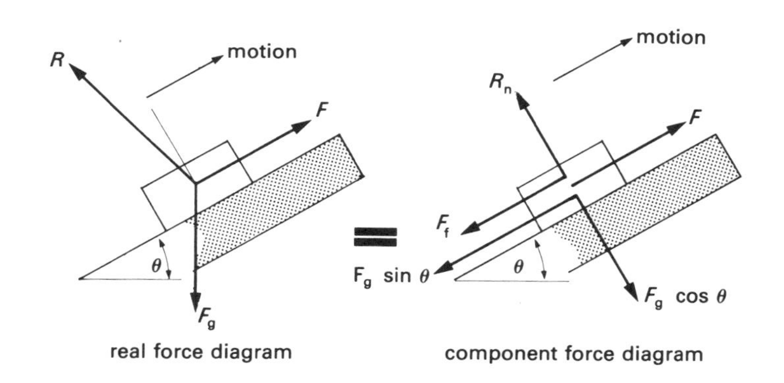

Diagram 5.29

Therefore for equilibrium, $R_n = F_g \cos \theta$

and $F_f + F_g \sin \theta = F$

Also, remember that $F_f = \mu R_n$

Example 13

Look at Example 6 again. A concrete slab of mass 50 kg is being pulled up a driveway by a rope parallel to the slope. The angle of the driveway is 20° and the coefficient of kinetic friction, μ_k, is 0·6. Find the force in the rope.

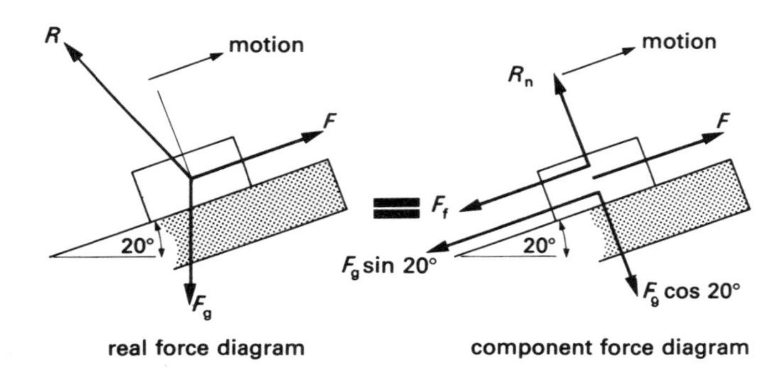

Diagram 5.30

Solution

See Diagram 5.30.

For equilibrium $R_n = F_g \cos 20$
$= 491 \times 0{\cdot}94$
$= 461$ N

but $F_f = \mu_k R_n$
$= 0{\cdot}6 \times 461$
$= 277$ N

Also for equilibrium, $F = F_f + F_g \sin 20$
$= 277 + (491 \times 0{\cdot}342)$
$= 277 + 168$
$= 445$ N

Therefore force in the rope is 445 N
(Note the slight difference to the graphical answer of 440 N.)

Examples 7, 8, 9 and 10 could also be done analytically, making the equilibrium equations suit the particular problem. The method for analytical solutions could be summarised as follows:

i Draw the real force diagram and from it construct the component force diagram.
ii Apply the equilibrium condition for forces perpendicular to the plane.
iii Remember $F_f = \mu R_n$.
iv Apply the equilibrium condition for forces parallel to the plane.
v From the 3 expressions, solve for the unknown quantity.

5.8 Perfectly smooth surfaces

In practice there really is no such thing as a perfectly smooth surface. If there was, then perpetual motion would be possible. However, there are times when a surface is so smooth that the resistance to bodies sliding along it is virtually negligible, i.e. the resistance is so small that it can be considered to be zero.

Whenever a surface is called 'perfectly smooth' or 'frictionless' then the surface reaction will be perpendicular to that surface.

Therefore R is R_n for perfectly smooth surfaces and F_f does not exist.

Example 14

A wooden beam leans at an angle of 60° to the ground and rests against a 'perfectly smooth' vertical wall. Find the angle of static friction between the ground and the bottom end of the beam. (Note that the mass of the beam is not important.)

Solution

In this case we must draw accurately a space diagram with angles and directions of forces. The information from the question is such that we can draw F_g from the mid-point of the beam and also we can draw the wall reaction horizontal at the top of the beam as shown in Diagram 5.31**a**. However, the beam is in equilibrium under the action of three forces.

i Gravitational force F_g, vertically down.
ii Wall reaction, horizontal.
iii Ground reaction, at some angle, at the bottom of the beam.

Therefore these three forces must either be parallel or concurrent. Since they obviously are not parallel, then their lines of action must all pass through a common point, the point of concurrency, as shown in

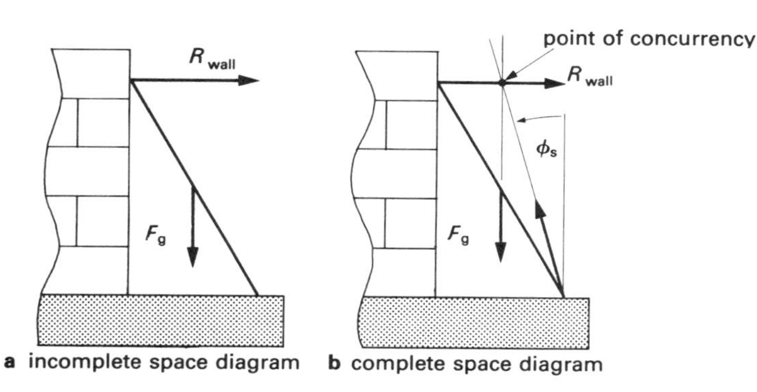

Diagram 5.31

Diagram 5.31**b**. Hence the direction of the ground reaction is obtained. The angle which the ground reaction makes to the vertical is the required angle of static friction and by measuring ϕ with a protractor, we find that $\phi = 16°$.

Therefore angle of static friction = 16°

5.9 Tractive resistance

This is the term used when wheels or rollers are the parts in contact with the surface. In such a case it is the resistance to rolling which is being considered and not a resistance to sliding.

Tractive resistance is the total resisting force and is measured in N. (Sometimes, it may be expressed in some books as the amount of resisting force per unit mass or weight of the body, e.g. N/kg, kN/tonne or N/kN, but this is now falling out of favour.) For our purposes, tractive resistance will always be the total resisting force to motion where wheels, rollers, etc., are involved. As a pure resisting force it will always act parallel to the surface and oppose the direction of motion (see Diagram 5.32).

5.10 Friction torque

When dealing with rotating parts such as clutches, brakes, bearings, etc., it is more useful to use friction torque, as a single item, instead of a friction force acting at some radius.

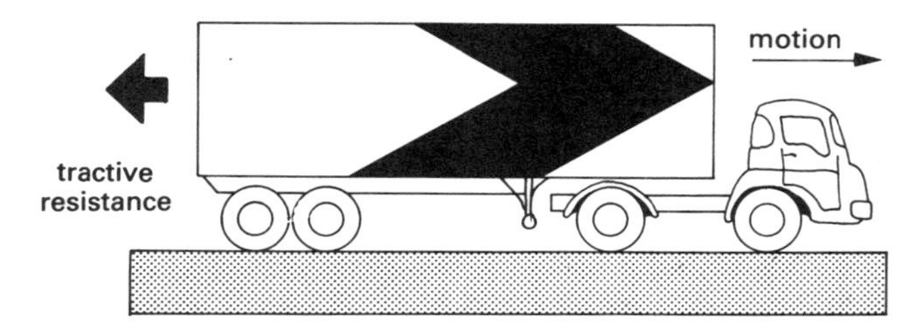

Diagram 5.32

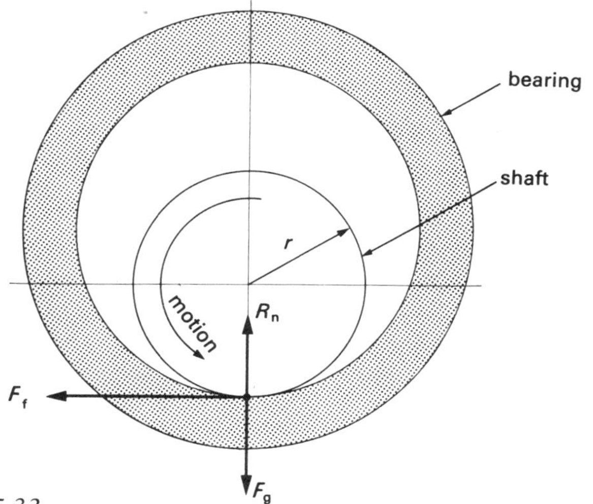

Diagram 5.33

Consider a shaft rotating in a bearing as shown in Diagram 5.33. Let the radius of the shaft be r metres and let the coefficient of kinetic friction be μ_K.

Diagram 5.33 has been drawn to illustrate the 'point contact' between shaft and bearing. At this point the surfaces in contact are really just very small horizontal surfaces with the upper body, the shaft, sliding to the right. Therefore the friction force F_f, which resists motion, acts horizontally to the left, i.e. tangential to the shaft. Therefore the resistance to the motion of the shaft expressed as a torque is:

friction torque = friction force × the radius at which it acts

therefore, $T_f = F_f \times r$ (in Nm)

but from previous work we know that $F_f = \mu R_n$

therefore friction torque, $T_f = \mu R_n$ r Nm

Example 15

A shaft weighing 1200 N is supported by bearings and the coefficient of kinetic friction is 0·15. If the shaft diameter is 180 mm, find the friction torque.

Solution

$$\mu_k = 0{\cdot}15$$
$$R_n = F_g = 1200 \text{ N}$$
$$r = \frac{0{\cdot}18 \text{ m}}{2}$$

therefore

$$T_f = \mu_k R_n r$$
$$= 0{\cdot}15 \times 1200 \times 0{\cdot}09$$

therefore friction torque = 16·2 Nm

5.11 Methods of reducing friction

In the early days of civilisation, man discovered that large heavy blocks of stone could be moved more easily if wooden rollers were used under the block. Since those early days, the science of friction has come a long way.

Modern methods of reducing friction effects are in four main categories:

1 Surface finish – by producing a smoother finish on the surfaces which will be in contact, e.g. surface grinding and polishing.

2 Materials in contact – by introducing fixed layers of material, e.g. bushes in bearings, varnishing wood, steel runners on sledges.

3 Lubrication – by using a wetting agent, e.g. water, oil, grease, depending on the other materials.

4 Inserting rollers – by changing from sliding friction to rolling, the resistance is very much lower, e.g. ball bearings, roller bearings, needle bearings, etc.

Exercises

1 Using Tan tables find the angle of friction for each of the following coefficients of friction: **a** 0·7, **b** 0·07; **c** 0·24; **d** 0·1; **e** 0·315.

2 Using Tan tables find the coefficient of friction for each of the following angles of friction: **a** 39°; **b** 27°; **c** 6·3°; **d** 21·8°; **e** 42°.

3 A sack of flour weighing 200 N is pushed along a bakery table. The limiting angle of static friction is 20° and the angle of kinetic friction is 15°. Find the horizontal force: **a** to start the sack moving, and **b** to keep the sack moving along the table.

4 A 100 kg crate is to be pushed into the corner of a garage. The coefficient of limiting static friction is 0·7 and the coefficient of kinetic friction is 0·5. Find the horizontal force: **a** to start the crate moving, and **b** to keep the crate moving along the floor.

5 A removal man pulls a tea-chest full of books across the floor of his van. He has to exert a horizontal pull of 200 N. The coefficient of kinetic friction is 0·2. Find the mass of the books in the tea-chest if the tea-chest has a mass of 2 kg when empty.

6 A horizontal force of 241 N was necessary to slide a wardrobe weighing 0·9 kN into place. Find the angle of friction and the coefficient of friction for the surfaces in contact.

7 A man finds that he can exert a maximum horizontal force of 648 N when he stands on a rough stone floor but only 408 N when on a wooden floor. His weight is 800 N.

a Explain the difference in the force he can exert.

b Find the angle of friction and the coefficient of friction in each case.

8 An empty dust-bin weighs 45 N and is pushed across a concrete floor by a long handle which is at 40° to the floor. Rubbish weighing 155 N is put into the dustbin and it is then pulled back across the floor by the same handle. The angle of kinetic friction is 20° for the surfaces in contact. Find the force in the handle: **a** to push the empty dustbin, and **b** to pull the loaded dustbin.

9 A boy, pushing a sledge along the ground, pushes down at an angle of 30° to the ground with a force of 132 N. The sledge and passenger together have a mass of 40 kg.

a Find the angle of friction for the surfaces in contact.

b If the boy had pulled up at an angle of 30° instead of pushing down, what force would he have used?

10 Find the smallest force (in magnitude and direction) which would make a mass of 10 kg just start to move along a horizontal surface. The limiting angle of static friction for the surfaces in contact is 18°.

11 A ramp at an angle of 15° is used for moving crates in and out of lorries. The average weight of the crates is 1·8 kN and the coefficient of friction is 0·4. Find the force parallel to the ramp to move a crate: **a** up the ramp and **b** down the ramp.

12 To clear his driveway, a car owner has to push a box weighing 200 N down a slope of 18°. He found that it took a force of only 20 N parallel to the slope. Find the angle of friction and the coefficient of friction for the surfaces in contact.

13 It takes a force of 12 N to make a box weighing 50 N start to move along a horizontal surface.

a Find the angle which the surface would have to make to the horizontal for the box to start sliding.

b At this angle of slope, what force would be necessary to make the box start moving upwards?

14 A body of mass 40 kg rests on a plane which makes an angle of 40° to the horizontal. The limiting angle of static friction is 25°. Find the least force parallel to the plane which will prevent the body sliding down.

15 A small machine of mass 100 kg is to be moved down from a platform 2 m high. Two ramps are available, one 4 metres long and the other 5 metres long, both having the same kind of surface. The coefficient of friction is 0·6. Which ramp should be used so that minimum force parallel to the ramp is required, and what is the minimum force?

16 A car pulls a trailer along a level road at constant speed. The tractive resistance for the trailer on the road is 250 N. Find the force in the draw-bar of the trailer.

17 A lorry weighing 30 kN coasts down a rough track at constant speed. If the tractive resistance is 3 kN, what is the slope of the track?

18 What braking-force must be applied to prevent an 850 kg car gaining speed when descending a hill of 1 in 8? Take the tractive resistance to be 640 N.

19 A train of total mass 40 tonnes travels down a uniform slope at constant speed without applying any driving force. The incline is 1 in 200. Find the tractive resistance.

20 What driving-force is required to overcome a tractive resistance of 500 N and accelerate a car of mass 1·25 tonnes from a speed of 20 km/h to a speed of 92 km/h in 10 seconds?

21 The caliper operated brake blocks on a bicycle were squeezed onto the rim of the 660 mm diameter wheel with a force of 20 N on each of the two blocks. The coefficient of friction is 0·6. Find **a** the total friction force acting at the rim, and **b** the friction torque caused by the blocks.

22 A torque of 0·6 Nm just causes a shaft to turn in its bearings. The shaft weighs 100 N and the coefficient of friction for the bearing materials is 0·15. Find the diameter of the shaft.

23 In a car starter motor, each of the two carbon brushes is held against the commutator by a spring-force of 2 N. The commutator is 30 mm in diameter and the coefficient of friction for the surfaces in contact is 0·15. Find the friction torque due to the brushes on the commutator.

24 On each wheel of a car two brake shoes act on a 200 mm diameter brake drum. The coefficient of friction may be taken to be 0·9. What force must be exerted on each brake-shoe to produce a friction torque of 450 Nm on each rotating road wheel?

Chapter 6

Work and Work Diagrams

6.1 Work

'Work' is a word which can have many different meanings in everyday use. It can mean the job we do, like digging the garden, painting, studying (as in 'homework') and many other things too. We might think that it is 'hard work' trying to move an obstinate boulder or trying to open a jammed window when all that is really happening is that we are becoming tired without having achieved anything. In engineering, 'work' has a particular meaning.

Work (W) is said to be done when a force moves the body on which it is acting in the direction of the force.

How then, can we measure work? Taking an empty trolley to the shops will obviously be easier to do than bringing the full trolley home again. The distance is the same, but the force needed on the homeward journey is larger.

Therefore the work done must depend on the size of the force. Also, it would be easier if we lived nearer to the shops, so the distance moved by the force must also affect the work done.

These two things (magnitude of the force and distance moved by the force) are the two things which we use to measure work.

Work = force × distance

We already know that force is measured in newtons (N), and distance is measured in metres (m). Therefore, when we multiply these two together we will get newton-metres (Nm). This unit of work is given a special name, the joule (symbol J).

W (in joules) = F (in newtons) × S (in metres)

$$W = F \times S$$

Therefore, 1 joule = 1 newton × 1 metre or, in words, '1 joule of work is done when a force of 1 newton moves the body on which it is acting 1 metre in the direction of the force'.

Example 1

It takes a horizontal force of 25 newtons to push a sledge along a level surface at uniform speed. Find the work done when the sledge has been pushed **a** 20 metres, **b** 50 metres and **c** 200 metres.

Solution

a Work = force × distance
$W = 25 \times 20$
$= 500$ Nm or J
therefore work = 500 J

b Work = force × distance
$W = 25 \times 50$
$= 1250$ Nm or J
therefore work = 1·25 kJ

c Work = force × distance
$W = 25 \times 200$
$= 5000$ Nm or J
therefore work = 5 kJ

Example 2

A barge is pulled along a canal at constant speed by a horse which walks along the tow-path. The rope from the horse to the barge is horizontal but makes an angle of 30° to the edge of the canal. The barge sails steadily down on the centre of the canal by using its rudder. The tension in the tow-rope is 200 N. Find the work done towing the barge a distance of 300 m.

Solution

The force which is doing the work is the force in the direction of the motion. Therefore we must find the component of the 200 N force which acts along the centre-line of the canal. (The component which is at right angles to the direction of motion is not doing any work.)

From the vector diagram in Diagram 6.1, the component of the

200 N force which acts along the centre-line of the canal is 173 N.

$$\begin{aligned} W &= \text{force} \times \text{distance} \\ &= 173 \times 300 \\ &= 51900 \text{ N m or J} \\ \text{therefore work} &= 51{\cdot}9 \text{ kJ} \end{aligned}$$

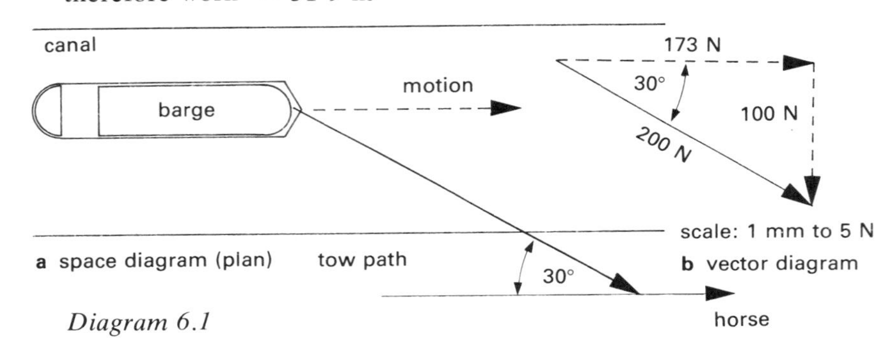

Diagram 6.1

6.2 Work diagrams

A very useful way of showing 'work done' clearly is to draw a graph. The horizontal axis will represent the distance moved, and the vertical axis will represent the force in the direction of motion.

Constant force

When the applied force is constant throughout the motion, the graph of force against distance will be a horizontal line as shown in Diagram

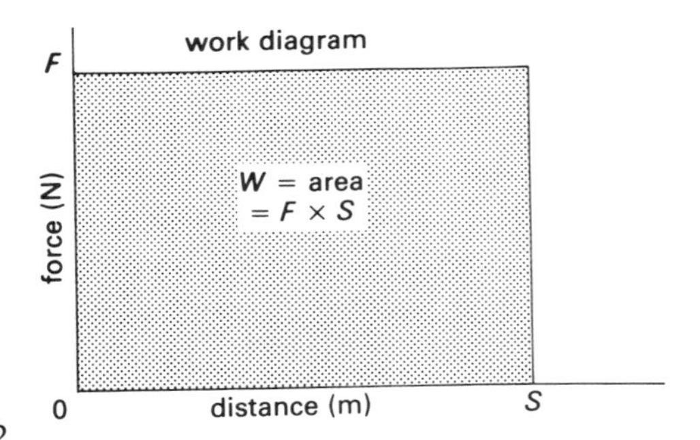

Diagram 6.2

6.2. This graph is called the **work diagram**. We know that $W = F \times S$, but on the work diagram (F × S) will give us the area of the shaded rectangle, i.e. the area under the graph.

Work = area of work diagram

Note that this is always true, even when the force is **not** constant throughout the motion.

Example 3

A bricklayer carried a load of bricks vertically up a ladder to a height of 15 m. The total weight of the bricks was 80 N. Draw the work diagram and hence find the work done in lifting the bricks.

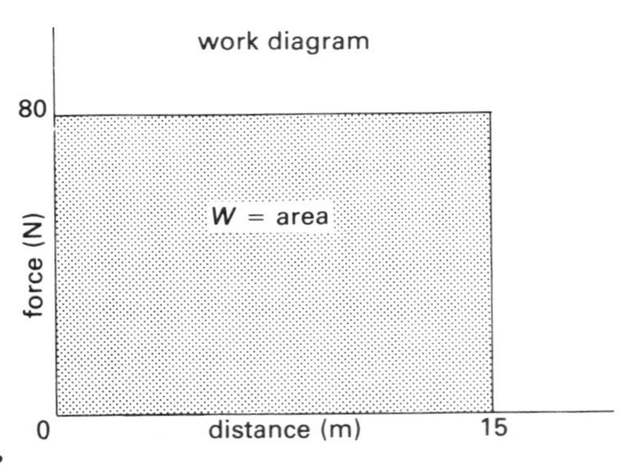

Diagram 6.3

Solution

From Diagram 6.3,

$$\begin{aligned} W &= \text{area of work diagram} \\ &= \text{area of shaded rectangle} \\ &= 80 \times 15 \\ &= 1200 \text{ N m or J} \\ \text{therefore work} &= 1{\cdot}2 \text{ kJ} \end{aligned}$$

Example 4

A man rolling his lawn pulled the roller against a tractive resistance of 60 N. When he finished, he had walked a total distance of 200 m. Draw the work diagram and find the work done.

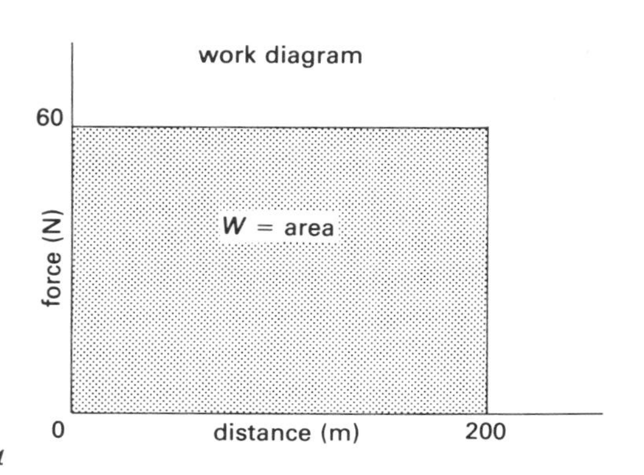

Diagram 6.4

Solution

Work = area of shaded rectangle in Diagram 6.4
= 60 × 200
= 12000 Nm or J
therefore work = 12 kJ

A changing force

Force can change in a number of ways:

i By varying all the time (Diagram 6.5**a**).

ii By changing from one constant force to another constant force (Diagram 6.5**b**).

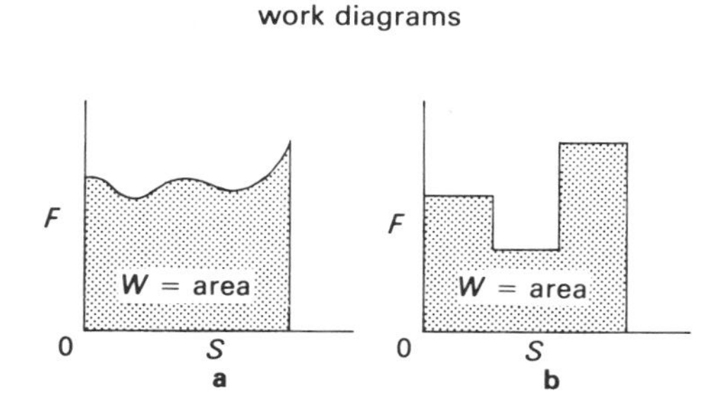

Diagram 6.5

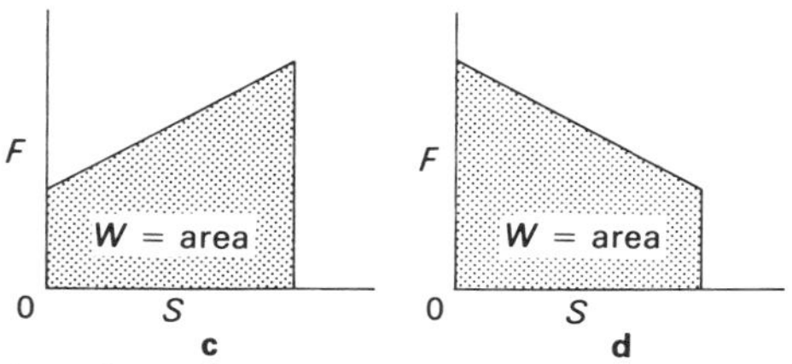

Diagram 6.5 (continued)

iii By increasing uniformly (Diagram 6.5**c**).

iv By decreasing uniformly (Diagram 6.5**d**).

v Or by any combination of these.

In our work, we will be dealing only with work diagrams which are made up of straight lines. Remember that work = area of work diagram in every case. Therefore it is important to draw the work diagram correctly.

Example 5

A lift in a block of flats started at the ground floor with 20 people in it whose average weight was 575 N. 10 people went out at the first floor and 10 at the second. The lift itself weighed 3·5 kN and the distance between floors was 4 metres. Draw the work diagram and find the total work done in raising the lift plus passengers to the second floor.

Solution

Before we can draw the work diagram, we must calculate the force at each stage.

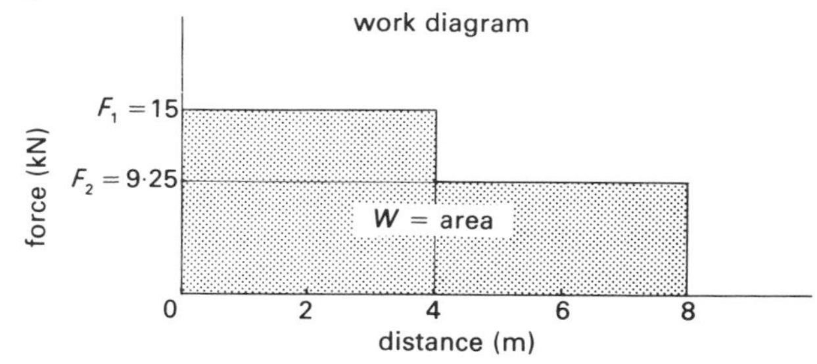

Diagram 6.6

At the ground floor, $F_1 = (20 \times 575) + (3{\cdot}5 \times 1000)$
$= 11500 + 3500$
$= 15$ kN
At the first floor, $F_2 = (10 \times 575) + 3500$
$= 5750 + 3500$
$= 9{\cdot}25$ kN
Therefore from Diagram 6.6,
work = area of work diagram
= sum of areas of rectangles
$= (15 \times 4) + (9{\cdot}25 \times 4)$
$= 60 + 37$
therefore work = 97 kJ

Example 6

A tension spring has a stiffness of 2 newtons per millimetre (i.e. to stretch it by 1 mm will take a force of 2 N, 10 mm will take 20 N, 100 mm will take 200 N and so on). A force is built up uniformly to 140 N on the spring. **i** Find how far the spring will have stretched. **ii** Draw the work diagram and find the work done in stretching the spring.

Solution

i To find how far the spring will have stretched we must use its stiffness, 2 N/mm of extension.

2 N stretches the spring 1 mm

therefore 140 N stretches the spring $\frac{140}{2} = 70$ mm

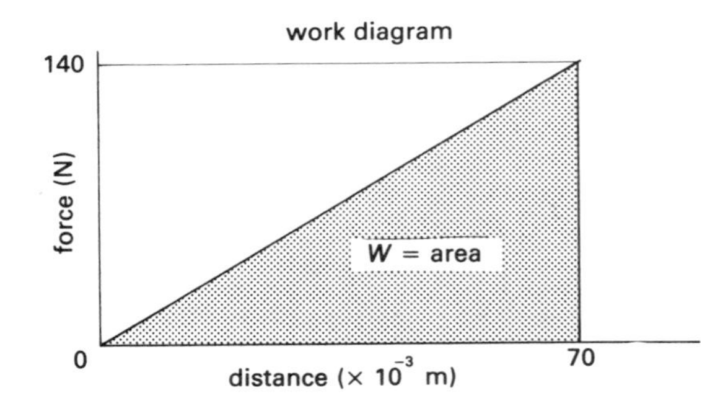

Diagram 6.7

ii To draw the work diagram, we must remember that initially the force is 0 and that it is increased uniformly to 140 N in stretching the spring 70 mm. Therefore, the graph must be as shown in Diagram 6.7 – an upward sloping straight line graph passing through the origin.

Again, work = area under the work diagram
= area of shaded triangle
$= \frac{1}{2} \times$ base $\times$ height
$= \frac{1}{2} \times (70 \times 10^{-3}) \times 140$
therefore work $= 4{\cdot}9$ J

Example 7

A 20 metre chain hangs down a well and has a weight of 15 N/m length. The chain is pulled up onto the edge of the well. Draw the work diagram and find the work done in raising the chain.

Solution

When the chain is first being lifted, the whole length has to be lifted. Therefore

initial force $= 20 \times 15$
$= 300$ N

Also, just as the last link of the chain slides over the lip of the well, there is no weight left to lift, therefore

final force $= 0$

Obviously, as we lift the chain, metre by metre, the force required gets less and less. Therefore the force is decreasing uniformly from

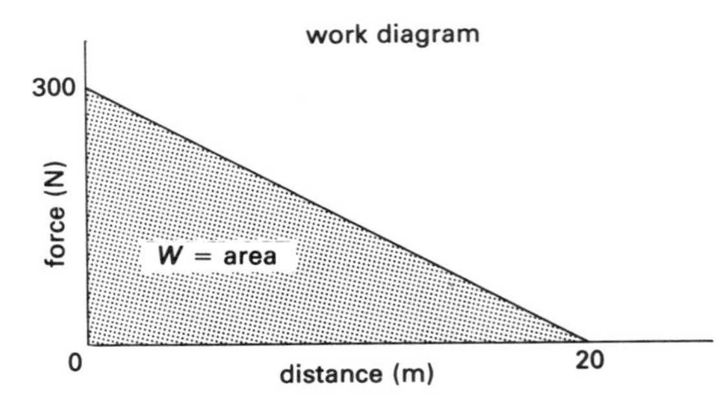

Diagram 6.8

300 N to 0 while lifting the 20 metre chain. The work diagram is as shown in Diagram 6.8.

From the work diagram,

$$W = \text{area of triangle}$$
$$= \tfrac{1}{2} \times 20 \times 300$$
$$\text{therefore work} = 3 \text{ kJ}$$

In Example 7 only a chain had to be lifted. If there was a bucket or some other object on the end of the chain then that would have to be lifted too. Look at Example 8.

Example 8

A log weighing 500 N had fallen down a well 25 metres deep and was to be lifted out using a strong chain. The chain had a weight of 16 N/m length. Draw the work diagram and find the work done raising both log and chain.

Solution

The force required can be considered to be made up of two parts; a constant force to lift the log and a uniformly decreasing force to lift the chain. Therefore,

$$\text{initial force} = \text{force to lift log} + \text{force to lift 25 m chain}$$
$$= 500 + (25 \times 16)$$
$$= 900 \text{ N}$$

Also, just at the instant that the log reaches the top,

$$\text{the final force} = (\text{force to lift log}) + 0 \text{ (no chain left)}$$
$$= 500 \text{ N}$$

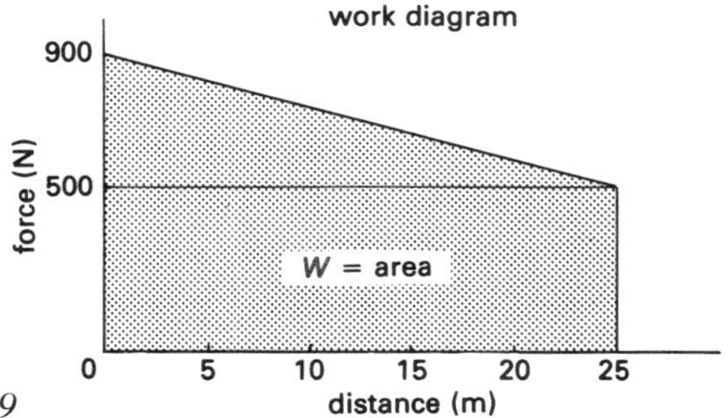

Diagram 6.9

We can now draw the work diagram as shown in Diagram 6.9. From the work diagram,

$$W = \text{area of work diagram}$$
$$= \text{area of rectangle} + \text{area of triangle}$$
$$= (500 \times 25) + (\tfrac{1}{2} \times 25 \times 400)$$
$$= 12500 + 5000$$
$$\text{therefore work} = 17{\cdot}5 \text{ kJ}$$

Note that the area of the shaded rectangle = W for the log alone and the area of the shaded triangle = W for the chain alone. So we could have said,

total $W = W_{\text{log}} + W_{\text{chain}}$

and dealt with the problem as two smaller problems.

Example 9

A boat drops anchor in 120 metres of water. The anchor, which has a mass of 500 kg, is lowered at constant speed using a chain of mass 5 kg per metre length. Draw the work diagram and find the work done in lowering the anchor to the sea bed.

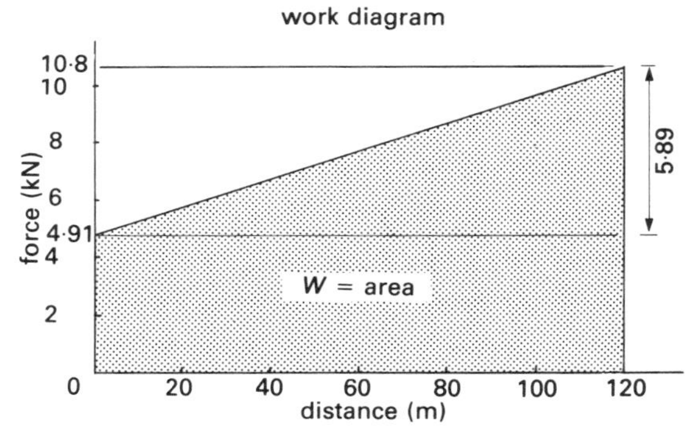

Diagram 6.10

Solution

$$\text{Initial force} = \text{anchor force} + \text{chain force}$$
$$= (500 \times 9{\cdot}81) + 0$$
$$= 4{\cdot}91 \text{ kN}$$

Final force = anchor force + chain force
= $(500 \times 9{\cdot}81) + (5 \times 120 \times 9{\cdot}81)$
= $4905 + 5886$
= $10{\cdot}8$ kN

From the work diagram (Diagram 6.10)
W = total shaded area
= area of rectangle + area of triangle
= $(4{\cdot}91 \times 120) + (\frac{1}{2} \times 120 \times 5{\cdot}89)$
= $589 + 353$
therefore $W = 942$ kJ

6.3 Pumps and work done by pumps

Very often liquids have to be moved along pipes from one point to another, or from one level to another. Most domestic central heating systems which use water radiators, need an electric water pump to make the hot water flow round the circuit. It is important to know how much water is to be pumped so that the size and strength of the pump can be calculated.

Consider a block of flats which is supplied with water from a tank on the roof. Let this storage tank be filled by a pump which pumps water from ground level. How could we find out how much work is done by the pump to fill the storage tank?

We must know certain things.

1 The volume of the tank (V) in m^3.

2 The density of water (ρ) in kg/m^3 (density is the mass of 1 m^3 of a substance).

3 The average height that the water has to be pumped (h) in metres.

The total work done by the pump will be the same as the work which would be done lifting the full tank from the ground up to the roof.

Total mass of water = volume × density
$m = V \times \rho$
therefore the weight of water = mg
= $V\rho g$
therefore work = force × distance
= weight of water × height
= $V\rho gh$
i.e. $W = V\rho gh$ (in this example)

(Note that the density of water = 1000 kg/m^3 = 1 kg/litre; 1 m^3 = 1000 litres and 1 litre = 1000 cm^3.)

Example 10

A tank of water is completely emptied twice a day, and is refilled each time by pumping water from a reservoir which is 10 m below the tank. The tank measures 3 m × 2 m × 2 m. Find the work done by the pump each day.

Solution

Volume of tank = $3 \times 2 \times 2$
= 12 m^3
Density of water = 1000 kg/m^3
therefore mass of water = 12×1000
= 12 000 kg
therefore weight of water = $12\,000 \times 9{\cdot}81$
= 118×10^3 N

To fill the tank once,
work = force × distance
= weight of water × height
= $118 \times 10^3 \times 10$
= $1{\cdot}18 \times 10^6$
= $1{\cdot}18$ MJ

therefore to fill the tank twice in a day,
work = $2{\cdot}36$ MJ

Example 11

A mine shaft 30 m deep is flooded with water to a depth of 6 m. The shaft has a cross-sectional area of 6 m^2. Find the work done in pumping water to ground level. Draw a work diagram.

Solution

Volume of water = 6×6
= 36 m^3
Density of water = 1000 kg/m^3
therefore mass of water = 36×1000
= 36×10^3 kg
therefore weight of water = $36 \times 10^3 \times 9{\cdot}81$
= 353×10^3 N

In this case the surface of the water is 24 m below ground level and the bottom of the shaft is 30 m below ground level. Therefore the average height for pumping is

$$\frac{30 + 24}{2} = 27 \text{ m}$$

work $= \text{force} \times \text{distance}$
$= \text{weight of water} \times \text{average height}$
$= 353 \times 10^3 \times 27$
$= 9{\cdot}53 \times 10^6$
therefore work $= 9{\cdot}53 \text{ MJ}$

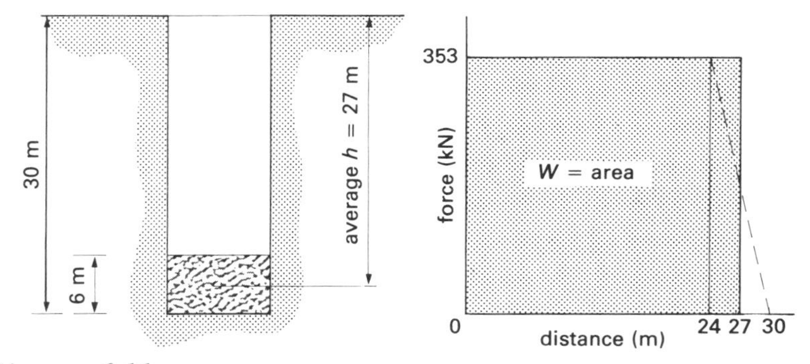

Diagram 6.11

There is another way of tackling this problem. We could consider the whole volume to be raised 'as a block' until the water started flowing over at ground level. Thereafter the lifting force would decrease uniformly to zero. This method would give a slightly different shape of work diagram, but of the same area (shown dotted in Diagram 6.11).

6.4 Work done against friction

We usually think of work being done in a useful way, but very often a lot of work is done just to overcome frictional resistance. In the last chapter, on 'friction', we saw that resistance to motion could be expressed in at least three ways:

i using a coefficient of friction ($F_f = \mu R_h$)
ii as a total tractive resistance
iii as a friction torque, for bearings, brakes, clutches, etc. ($T_f = \mu R_n r$)

Wherever there is friction, work will be done against friction when the body moves.

Example 12

A workshop bench was pushed 3 m across the floor. The coefficient of friction for the surfaces in contact was 0·6 and the bench had a mass of 80 kg. How much work was done against friction?

Solution

We already know the distance moved but we need to find the force of friction.

$F_f = \mu R_n$
$= 0{\cdot}6 \times 80 \times 9{\cdot}81$
$= 471 \text{ N}$
therefore work $= \text{force} \times \text{distance}$
$W = 471 \times 3$
$= 1413 \text{ Nm or J}$
therefore work done against friction $= 1{\cdot}41 \text{ kJ}$

6.5 Work done by a torque

Diagram 6.12 shows a shaft of radius r with a force F acting tangential to the shaft. The torque on the shaft is $T = F \times r$.

Let the applied force F make the shaft turn one revolution. Therefore F will have travelled once round the circumference of the shaft.

Distance moved by $F = \text{circumference of shaft}$
$= 2\pi r$
therefore work done $W = \text{force} \times \text{distance}$
$= F \times 2\pi r$
$= 2\pi F r$
But torque $T = Fr$
therefore $W = 2\pi T$ for 1 revolution
Therefore for any number of revolutions, say N,

the total work done = work done for 1 revolution × number of revolutions
therefore $W = 2\pi N T$

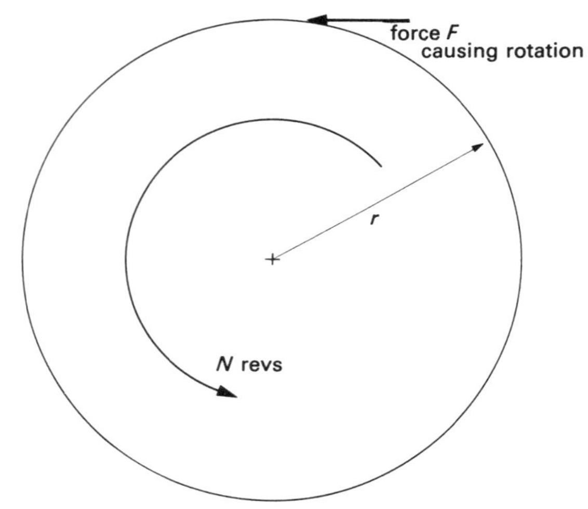

Diagram 6.12

Example 13

A drive-shaft transmits a torque of 10 N m and makes a total of 200 revolutions. Find the total work done by the torque.

Solution

For a torque $W = 2\pi NT$
$= 2\pi \times 200 \times 10$
$= 12{\cdot}6$ kJ

therefore total work done $= 12{\cdot}6$ kJ

Example 14

A shaft weighing 60 N is supported by two identical bearings, one at each end of the shaft. The shaft has a diameter of 50 mm and the coefficient of friction for the bearing surfaces in contact is 0·2. For 20 minutes the shaft runs at a speed of 500 rev/min. Find the total work done against friction in the bearings.

Solution

Total friction torque, $T_f = 2 \times$ friction torque in each bearing
$= 2 \times (\mu \times R_n \times r)$

For each bearing, $R_n = \frac{60}{2}$ N

therefore $T_f = 2 \times (0{\cdot}2 \times \frac{60}{2} \times 0{\cdot}025)$
$= 2 \times 0{\cdot}15$
$= 0{\cdot}3$ N m

Also, total number of revolutions $N =$ rev/min $\times$ number of minutes
$= 500 \times 20$
$= 10\,000$ rev

therefore total work done $W = 2\pi N T_f$
$= 2\pi \times 10\,000 \times 0{\cdot}3$
$= 18{\cdot}8 \times 10^3$
$= 18{\cdot}8$ kJ

Therefore total work done against friction in the bearings $= 18{\cdot}8$ kJ

6.6 Work done to accelerate a body

In order to cause acceleration, we know that a force greater than the resistance to motion is required. The total work done will be equal to the total force times the total distance moved.

Total $W =$ total $F \times S$

But the total force is made up of at least two parts:

total $F =$ force to overcome friction + force to produce acceleration
$= F_f + ma$

therefore total $W = (F_f + ma) \times S$

Example 15

A curling stone is to be accelerated at 0·5 m/s² over a distance of 2 m. The curling stone has a mass of 19·3 kg and the coefficient of friction for the surfaces in contact is 0·15. Find **i** the required total force, and **ii** the work done.

Solution

i Total force F = friction force + accelerating force
$= F_f + ma$
$= \mu R_n + ma$
$= \mu mg + ma$
$= (0{\cdot}15 \times 19{\cdot}3 \times 9{\cdot}81) + (19{\cdot}3 \times 0{\cdot}5)$
$= 28{\cdot}4 + 9{\cdot}65$
$= 38{\cdot}1$ N

ii Work done $W = F \times S$
$= 38{\cdot}1 \times 2$
$= 76{\cdot}2$ J

Exercises

1 Find the work done when a force of 200 N moves its point of application a distance of 3·5 m.

2 25 kJ of work was done raising a crate a height of 5·1 m. Find the mass of the crate.

3 During one day a man did 240 kJ of work moving a trolley up and down passageways in a warehouse. The average force which he had to exert was 160 N. How far did he move the trolley in one day?

4 A boy does 40 kJ of work pushing a sledge a distance of 800 m. Find the force he exerted while pushing the sledge.

5 A railway truck is moved along a level track by a rope which makes a horizontal angle of 30° to the side of the track. The tension in the rope is 716 N. Find the work done moving the truck 150 m.

6 A builder's hoist raises 200 bricks a height of 4 m. At this level, 80 bricks are unloaded and the remainder are taken 4 m higher to the top level where the hoist is emptied. Each brick weighs 35 N. Draw the work diagram and find the work done raising the bricks.

7 A delivery boy uses a sledge which weighs 80 N to deliver two orders. The first order weighs 200 N and has to go to a house 500 m from the shop. The second order weighs 120 N and has to go 250 m further than the first. After making the deliveries the boy returns to the shop by the same route. Draw the work diagram and find the work done against friction for the round trip. (Take μ = 0·4.)

8 A spring is stretched 50 mm by a 76 N force. Find **i** the stiffness of the spring, and **ii** the work done in stretching the spring.

9 A spring has a stiffness of 12 N/mm of compression and is compressed 20 mm by a load. Find **i** the load (in N), and **ii** the work done in compressing the spring.

10 A chain 15 metres long, weighing 20 N/m length, hangs vertically from a drum. Find the work done in winding the chain onto the drum.

11 A lift in a 60 m high block of flats has a mass of 300 kg. It is raised by four cables each of which has a mass of 0·5 kg/m length. Find **i** the work done raising the empty lift to the top flat and **ii** the work done raising the empty lift to the middle flat.

12 A well is 1·2 m in diameter and the water level is 20 m below ground level. The water is 4 m deep and has a density of 1000 kg/m³. Find the work done emptying the well.

13 A reservoir supplies 60 m³ of water a minute to a hydro-electric station 100 m below the level of the reservoir. Find the work (per minute) which could be done by the water supply.

14 Crude oil from a tanker is to be pumped up to storage tanks on the dockside. The average height that the oil has to be raised is 30 m and the oil has a density of 800 kg/m³. The storage tanks are 25 m in diameter and 20 m high. **i** Find the work done pumping enough oil to fill one tank. **ii** The tanker is to deliver 1·2 million barrels of oil. How many tanks will be needed. (1 barrel = 159 litres.)

15 A man pushed his car 0·8 km along a level road to a garage. The work he did was 96 kJ. Find the tractive resistance.

16 A horizontal conveyor belt moves 10 packages each having a mass of 1·02 kg a distance of 5 m. The coefficient of friction between the belt and its supporting surface is 0·15. Find the work done moving the packages.

17 An injured rock-climber is pulled up a 60° slope on a stretcher. The combined mass is 775 kg, the length of the slope is 60 m and the coefficient of friction is 0·4. **i** Find the work done. **ii** For a straight vertical lift, what would be the work done? Explain the difference between this result and the result for part **i**.

18 A winch does 20 kJ of work hauling a 150 kg casting 40 m across a workshop floor. Find the coefficient of friction for the surfaces in contact.

19 The rotor of a turbine has a mass of 120 kg and a shaft diameter of 75 mm. It runs in bearings, where the coefficient of friction is 0·01, at a uniform speed of 4000 rev/min. Find the work done per minute against friction in the bearings.

20 A motor-cycle brake does 166 J of work for 4 revolutions of the wheel. The brake drum has a diameter of 220 mm and the coefficient of friction is 0·6. Find the force exerted by the brake on the brake drum.

Aerial view of part of the Conon Valley hydroelectric scheme

Chapter 7

Energy and Power

7.1 Energy

Have you ever felt that you had no energy, or that you did not have enough energy to do something? I'm sure you have heard of certain foods or drinks which are supposed to be particularly good for giving energy very quickly.

When we feel we are 'full of energy' we feel that we can go out and do a 'hard day's work'. The amount of work we can do will depend on how much energy we have. It is exactly the same in engineering, and we define energy as follows.

Energy is the capacity to do work

This means that there must be a store of energy and that work is drawn from the store. If we could measure the store of energy we could say that it would provide that amount of work.

This is rather like the water storage tank in a house. It holds a certain amount of water and we can draw water from the tank as and when we need. In our bodies we store energy by eating food and this enables us to do work when we need to.

The water tank can be filled up again when it is empty and we can eat more food when our energy level gets low. We can replace the energy supply.

In engineering, both work and energy are measured in joules (J), kilojoules (kJ), megajoules (MJ) etc. The symbol we use for energy is E.

7.2 Forms of energy

Energy is available in a variety of forms and it can be stored in a variety of ways.

1 Strain energy (E_s) is the energy which is stored in a body because it is stretched or compressed. The work which is done in compressing a spring is stored in the compressed spring as strain energy – a capacity to do work. When the spring is released it will do work as it expands back to its original size and shape. For example, clockwork motors, clocks, catapults, spring-boards, etc. Therefore:

$$\begin{aligned}\text{strain energy} &= \text{work done in compressing the spring}\\ &= \tfrac{1}{2}Fx\end{aligned}$$

(where x is the distance that the spring has been compressed: see Chapter 6). Therefore:

$$E_s = \tfrac{1}{2}Fx$$

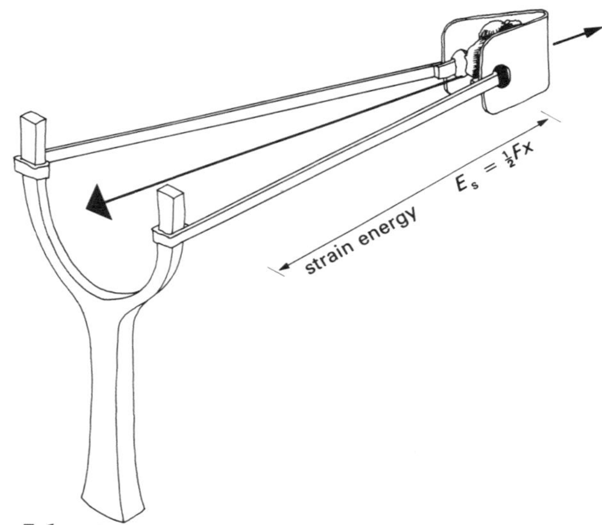

Diagram 7.1

2 Potential energy (E_p) is the energy which a body possesses because of its position, i.e. the vertical height through which the body could fall.

The work which is done in lifting a mass (m) to a height (h) is stored in the body as potential energy and if the body is allowed to fall, it could do work as it returned to its original position. For example, pile drivers, weights on old fashioned wall-clocks, water in a reservoir for a hydro-electric station etc.

Therefore:

potential energy = work done in lifting a mass

$= F_g \times h$

$= mgh$

therefore $E_p = mgh$

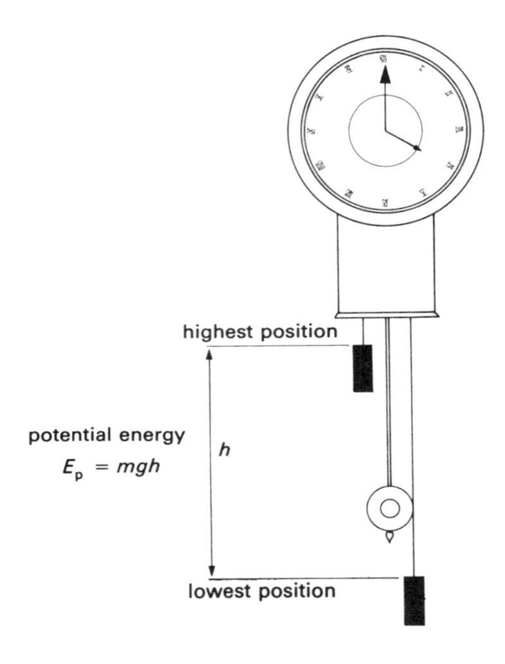

Diagram 7.2

3 Kinetic energy (E_k) is the energy which a body has because of its speed. The work which is done, accelerating a body from rest up to a certain speed, is stored in the moving body as kinetic energy. The motion of the body can then be used to do work and the body will slow down. When the body is at rest again, all its kinetic energy will have been used up doing work; for example, flywheels on engines, (this is a very important use of kinetic energy).

Consider a cyclist who pedals along a level road and then stops pedalling. Gradually he slows down until eventually he comes to a complete halt. All of his kinetic energy has been used up in doing work against the air resistance and against the tractive resistance.

Consider a car which has to stop at traffic lights. The driver puts on the brakes until the car comes to a halt. All the kinetic energy of the car

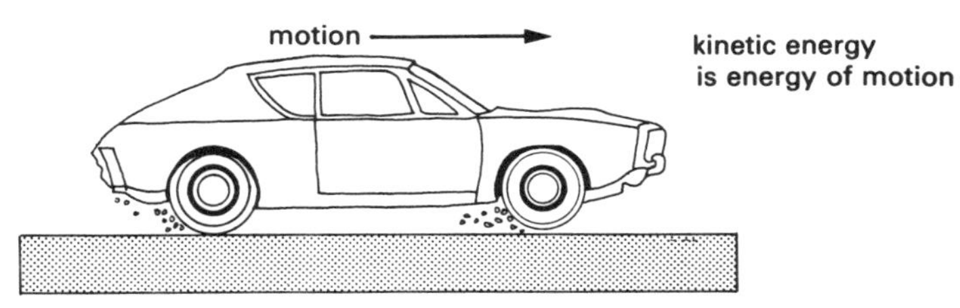

Diagram 7.3

has been used up doing work against the frictional resistance of the brakes.

Kinetic energy = work which could be done by a body in coming to rest

4 Electrical energy can occur naturally, e.g. lightning. It is much more common for electrical energy to be produced by generators or dynamos and this energy can be used for doing work in a very wide range of situations. For example, electric fires, lamp bulbs, electromagnets, electroplating, electrolysis, hi-fi equipment, motors, arc-welding, etc. Storing electrical energy is not as common as you might think. A battery is not really a store of electrical energy, it is a store of chemical energy which changes to electrical energy when you need it. Electrical energy is usually produced on demand, either by generators, dynamos or batteries, but it can be stored in a capacitor.

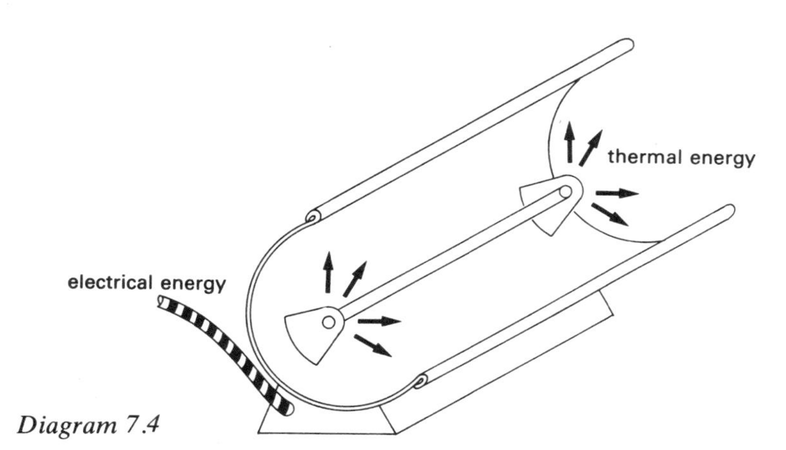

Diagram 7.4

5 Thermal energy is heat energy. We can get this from the sun, by burning fuel, by passing an electric current through an electric fire, by causing friction, etc. It can be stored, for example, in storage heaters which are quite popular for house-heating because they can be used at off-peak times when the electrical energy is cheaper. The heat produced is then partly used immediately for heating the house and partly stored for use later when the off-peak supply is cut off. About half of the heat produced is stored in the storage heaters for use later.

6 Chemical energy is the energy a body contains because of the substances it is made of. Our digestive system releases this energy from the food we eat, but not completely. For example, an average eating apple contains enough energy to do about 170 kJ of work (i.e. the work which would be done lifting a family-size car 15 to 20 metres into the air). This is the work which could be done, if we could get all the energy out of the apple. Batteries have already been mentioned as a way of storing energy chemically but the most important chemical energy, for our purposes, is the energy contained in fuel – coal, gas, petrol, oil, etc. The energy contained by a fuel can be measured so that we know how much energy will be released by the fuel when it is burned. For example, burning the following fuels would release the amount of energy shown:

coal: 22 to 35 MJ/kg
petrol: 42 to 48 MJ/l
natural gas: 36 to 40 MJ/m^3

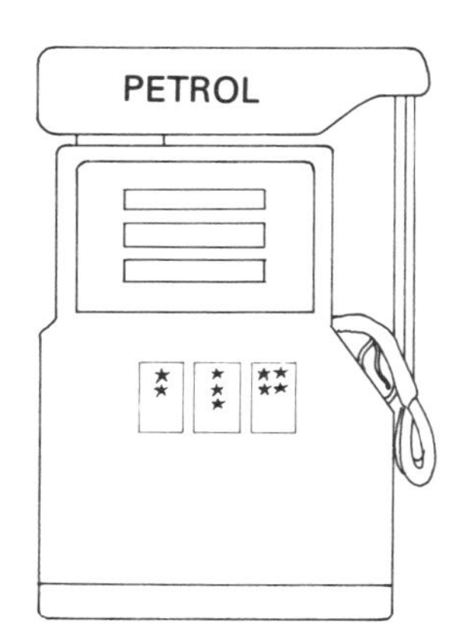

Diagram 7.5

(Note that the energy is usually given in MJ/kg for a solid fuel, MJ/l for a liquid fuel and MJ/m³ for a gaseous fuel.)

7.3 Change of form of energy

Energy does not always stay in one form, it can be changed from one form to another. Some examples of the common changes of form have already been mentioned in the passing. Let us look at a few more examples.

i Brakes

All brakes are designed to do a lot of work very quickly. The high friction forces create a lot of heat which is usually given off to the surrounding air. The energy transfer is:

kinetic energy → thermal energy

ii Springs

When work is done on a spring, it is stored as strain energy:

work → strain energy

When the spring is allowed to return to its original size and shape, it will do work:

strain energy → work

iii Fires

a In electric fires:

electrical energy → thermal energy

b In gas, coal or oil fires:

chemical energy → thermal energy

iv Motors/generators

a Electric motors:

electrical energy → work

b Generator:

work → electrical energy

v Batteries

a Discharging:

chemical energy → electrical energy

b Charging:

electrical energy → chemical energy

vi Photo-electric cells

light → electrical energy

vii Thermocouples

thermal energy → electrical energy

viii Pile-drivers

a Being raised:

work → potential energy

b Being dropped:

potential energy → kinetic energy → work

ix Car engines

Chemical energy → thermal energy → work (on piston) → kinetic energy (at flywheel)

In a car engine, the chemical energy of the fuel is changed into thermal energy when the fuel is burned. This makes the gases expand and push the piston which in turn accelerates the crankshaft and flywheel. The surplus kinetic energy at the flywheel is then available to do work to accelerate the car.

x Steam turbogenerator

Chemical energy of fuel → thermal energy of steam → kinetic energy of turbine → electrical energy output from generator

xi Hydroelectric generator

Diagram 7.7 shows an outline of how an impulse turbine could be set up. (The arrangement for a reaction turbine would be slightly different.)

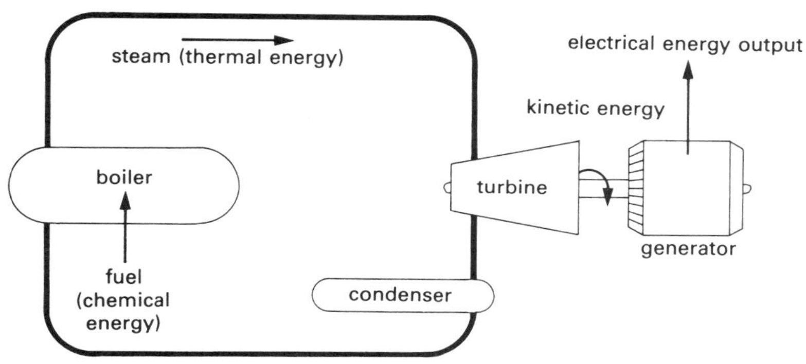

Diagram 7.6

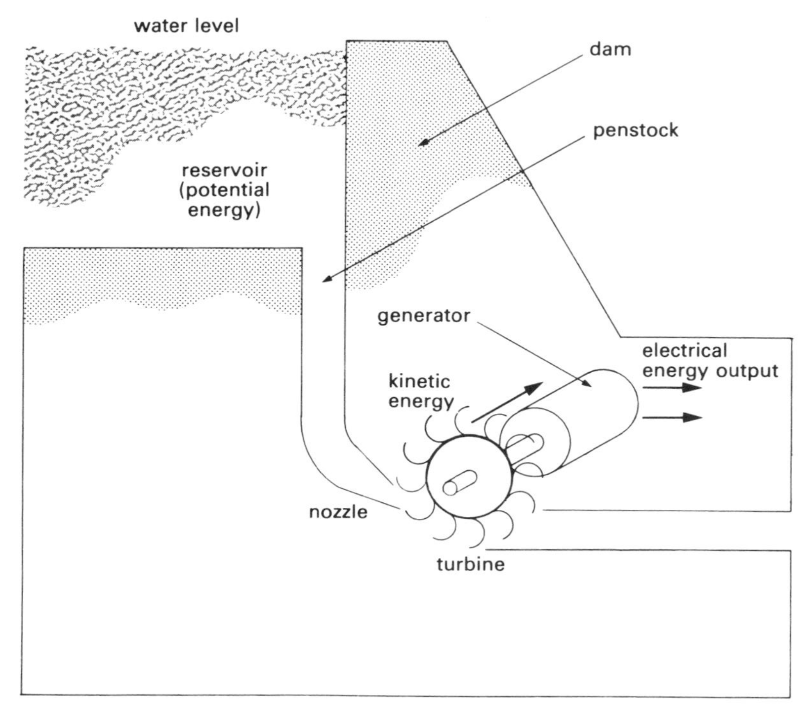

Diagram 7.7

Potential energy of water → kinetic energy of water → kinetic energy of turbine → electrical energy output from generator

At off-peak times the demand for electricity is quite low. In some hydroelectric stations the surplus electrical energy from other stations (via the National Grid) is used to pump water back up to the reservoir for use when peak demand time comes round again. These stations, which are hydroelectric pumped storage schemes, allow other electricity generating stations to operate at their most efficient level.

electrical energy → work → potential energy

xii Mine shaft hoisting gear

a Raising the lift:

electrical energy → work → potential energy

b When lowering the lift, it is not simply allowed to fall, but is controlled as it descends by making the motor run as a generator, thus producing useful electrical energy again:

potential energy → work → re-generated electrical energy

(This is sometimes called 'regenerative braking'.)

7.4 Conservation of energy

We have looked at many ways of changing energy from one form to another without considering how much energy has been usefully changed or how much energy has been wasted in the process. The energy has always come from somewhere, has been changed and has gone out in a different form. It is not possible to 'manufacture' or create energy, it has to be collected. Also, it is not possible to lose or destroy energy, it must go somewhere.

The principle of conservation of energy states that energy can neither be created nor destroyed. It can only be changed from one form to another. Therefore, at the end of a change of energy, if we seem to have less energy than we started with, it means that some energy has been lost in friction, heat, light, sound etc.

In other words, energy which is lost is energy which has been changed into a form which was not the intended change. We can write this as an equation:

energy input = energy output + energy lost

Example 1

100 kJ of electrical energy was supplied to an electric motor driving a winch which raised a load of 612 kg a height of 10 m. Find **i** the energy output, i.e. work done and **ii** the energy lost. State where the lost energy could have gone.

Solution

i Energy output = work done by winch raising load
$= 612 \times 9{\cdot}81 \times 10$
$= 60 \times 10^3$ J
$\therefore$ energy output = 60 kJ

ii Energy lost = energy input − energy output
$= 100 - 60$
$\therefore$ energy lost = 40 kJ

This could have been lost in a number of ways: friction losses, noise, sparking in the motor, heating the motor etc.

Example 2

A car engine produced 78 MJ of work when it used 5 l of fuel. 1 l of the fuel contained 48 MJ of chemical energy. Find **i** the total energy input and **ii** the total energy lost.

Solution

i Energy input = energy per litre × number of litres
$= 48 \times 5$
$\therefore$ energy input = 240 MJ

ii Energy lost = energy input − energy output
$= 240 - 78$
$\therefore$ energy lost = 162 MJ

7.5 Efficiency

When energy is changed from one form to another, there is always some energy lost. We can measure the energy input and the energy output, and from these two quantities we can tell how well the change has been done. This is done by working out the fraction of the energy input which has successfully been changed to the output required. This is called the **efficiency** of the change. Efficiency (η) is the ratio of the energy output to the energy input.

$$\text{efficiency } (\eta) = \frac{\text{energy output}}{\text{energy input}}$$

The symbol for efficiency, η, is the Greek letter eta. Efficiency is usually stated as a decimal fraction, e.g. 0·95, 0·6, 0·28 etc., but it could equally well be given as a percentage,

e.g. 0·95 = 95%
0·6 = 60%
0·28 = 28% and so on.

In Example 1, efficiency, $\eta = \dfrac{\text{energy output}}{\text{energy input}} = \dfrac{60}{100}$

$\therefore \eta = 0{\cdot}6$ or 60%

In Example 2, efficiency, $\eta = \dfrac{\text{energy output}}{\text{energy input}} = \dfrac{78}{260}$

$\therefore \eta = 0{\cdot}3$ or 30%

Example 3

Find the efficiency of a process which changes an energy input of 120 MJ to a useful energy output of 54 MJ.

Solution

Efficiency, $\eta = \dfrac{\text{energy output}}{\text{energy input}} = \dfrac{54}{120}$

$\therefore \eta = 0{\cdot}45$ or 45%

Example 4

A process is 85% efficient. Find the useful energy output when the energy input is 52 MJ.

Solution

Efficiency, $\eta = \dfrac{\text{energy output}}{\text{energy input}}$

$\therefore$ energy output $= \eta \times$ energy input

$$= 85\% \text{ of } 52$$
$$= 0{\cdot}85 \times 52$$
$$\therefore \text{energy output} = 44{\cdot}2 \text{ MJ}$$

Example 5

A machine has to do 14 MJ of work and it is known that the machine is 0·35 efficient. Find how much energy must be supplied to the machine so that it will do the necessary work.

Solution

$$\text{Efficiency,} \quad \eta = \frac{\text{energy output}}{\text{energy input}}$$
$$\therefore \text{energy input} = \frac{\text{energy output}}{\eta}$$
$$= \frac{14}{0{\cdot}35}$$
$$\therefore \text{energy input} = 40 \text{ MJ}$$

7.6 Power

We have dealt with work and energy and have considered energy changes and efficiency. The one thing we have not yet considered is the time taken for work to be done or an energy change to be completed. Unloading a shipment of coal could be done by one man using a shovel and a barrow, but it would take him a long time to do the work. If a machine was used it could do the work more quickly. The work which is done is the same, but the time taken to do the work is different. We say that the one which does the work more quickly is more powerful. **Power** (P) is the rate of doing work or the rate of transfer of energy.

$$\text{Power } (P) = \frac{\text{work done}}{\text{time taken}} = \frac{\text{energy transfer}}{\text{time taken}}$$

The units of power must therefore be joules per second (J/s), kilojoules per second (kJ/s) etc. The 'joule per second' is given a special name, the watt (W).

1 J/s = 1 W

As with other units, we can have multiples and sub-multiples of the watt, e.g. mW, kW, MW, GW.

Examples of power ratings

It may help us to get an idea of what 1 W is or what 1 kW is if we look at some examples of devices designed to change energy from one form to another. Remember that the larger the power rating the more energy is transferred in the same time.

Perhaps the most common examples are in and around the home:

electric light bulbs: 45 W, 60 W, 100 W, 150 W
a one bar electric fire: 1 kW
a fast-boiling electric kettle: 2·5 to 3 kW
the engine in a family car: 45 to 60 kW (this is equivalent to 60 to 80 horsepower which is the unit of power still being used by car manufacturers).

In industry some very powerful machines are used and therefore the Electricity Boards must be able to provide a suitable supply. An electricity power station may have several turbogenerator units producing electricity. A large steam turbogenerator unit, running at a speed of 3000 revolutions per minute, would generate 660 MW of electrical power.

The largest hydroelectric power plant in the world is in Russia on the Yenisey River close to Krasnoyarsk. It has a capacity of 6000 MW. For the space programme, very powerful rocket engines have been developed. A rocket engine can produce more power for its size than any other kind of engine.

7.7 Calculating power

We have already seen that power is the rate of doing work or the rate of transferring energy, and we know that work is calculated by force × distance.

$$\text{Power} = \frac{\text{work done}}{\text{time taken}}$$
$$\therefore \text{power} = \frac{\text{force} \times \text{distance}}{\text{time taken}}$$

Therefore, writing these expressions in symbols, we have:

$$P = \frac{W}{t} \text{ or, } P = \frac{F \times S}{t}$$

Example 6

Find the average power of a man who raises a 51 kg paving slab a height of 2 m in 4 s. (250 W)

Solution

$$\text{Average power} = \frac{\text{total work done}}{\text{time taken}} = \frac{\text{force} \times \text{distance}}{\text{time taken}} = \frac{m\,g\,h}{t} = \frac{51 \times 9{\cdot}81 \times 2}{4}$$

$\therefore$ average power = 250 W or 0·25 kW

Example 7

An electric motor which is rated at 5 kW, has to do 160 kJ of work. Find the time it will take the motor to do the work. (32 s)

Solution

$$\text{Average power} = \frac{\text{work done}}{\text{time taken}}$$

$$\therefore \text{time taken} = \frac{\text{work done}}{\text{average power}} = \frac{160 \times 10^3}{5 \times 10^3}$$

$\therefore$ time taken = 32 s

Example 8

A 2·5 kW pump is used to pump water from a well which is 10 m deep. How many litres of water per minute can the pump deliver? (1530 l/min)

Solution

$$\text{Average power} = \frac{\text{work done}}{\text{time taken}}$$

$\therefore$ work done = average power $\times$ time taken
= $2{\cdot}5 \times 10^3 \times 60$
= 150×10^3 J

but, work done = weight of water $\times$ height
= $(m \times g) \times h$
= $m\,g\,h$

$\therefore 150 \times 10^3 = m \times 9{\cdot}81 \times 10$

$$\therefore m = \frac{150 \times 10^3}{9{\cdot}81 \times 10} = 1530 \text{ kg}$$

But, 1 litre of water has a mass of 1 kg

$\therefore$ pump can deliver 1530 l of water per minute

Relationship between power and speed

$$\text{Average power} = \frac{\text{work done}}{\text{time taken}} = \frac{\text{force} \times \text{distance}}{\text{time taken}}$$

From the definition of average speed,

$$\text{average speed} = \frac{\text{distance travelled}}{\text{time taken}}$$

therefore average power = force $\times$ average speed

In general, power = force $\times$ speed

or $P = F \times v$

Example 9

A force of 120 N moves a body at a uniform speed of 5 m/s. Find the power. (600 W)

Solution

Power = force $\times$ speed
= 120×5

$\therefore$ power = 600 W

Example 10

A boy uses a power of 100 W to push a sledge at a speed of 2 m/s. What force did he exert? (50 N)

Solution

Power = force $\times$ speed

$$\therefore \text{force} = \frac{\text{power}}{\text{speed}} = \frac{100}{2}$$

$\therefore$ force = 50 N

Relationship between power and torque

In Chapter 6, we saw that the work done by a torque was given by the equation:

$$W = 2\pi NT$$

where N is the total number of revolutions.

$$\therefore \text{power} = \frac{\text{work done}}{\text{time taken}}$$

$$\therefore P = \frac{2\pi NT}{t}$$

Usually, the speed of rotation is given as revolutions per second (rev/s) or revolutions per minute (rev/min). The symbol for speed of rotation is n. Therefore, where n is in rev/s, the work done per second is given by:

work done per second $= 2\pi nT$

But this is the power.

$\therefore$ power, $P = 2\pi nT$ (where n is in rev/s)

Example 11

An electric motor has a power-rating of 0·4 kW and a running speed of 1380 rev/min. Find **i** the torque available at motor spindle and **ii** the force which would be transmitted at the rim of a 30 mm diameter pulley fitted to the spindle. (2·77 Nm, 185 N)

Solution

i Power, $P = 2\pi nT$

torque $T = \frac{P}{2\pi n}$

$= \frac{0{\cdot}4 \times 10^3 \times 60}{2\pi \times 1380} \quad (n = \frac{1380}{60}\text{ rev/s})$

torque $T = 2{\cdot}77$ N m

ii Torque $=$ force $\times$ radius

force $= \frac{\text{torque}}{\text{radius}}$

$= \frac{2{\cdot}77}{15 \times 10^{-3}}$

force $= 185$ N

Example 12

A motor driven winch has a winding drum 200 mm in diameter which rotates at 30 rev/min while exerting a force of 2·5 kN on the winch cable. Find **i** the power output of the winch, and **ii** the power input for the winch if its efficiency is 0·8. (785 W, 981 W)

Solution

i Torque on drum $=$ cable force $\times$ drum radius

$= (2{\cdot}5 \times 10^3) \times (100 \times 10^{-3})$

$\therefore T = 250$ Nm

$\therefore$ power output $= 2\pi nT$

$= 2 \times \pi \times \frac{30}{60} \times 250$

$\therefore$ power output $= 785$ W

ii Efficiency $= \frac{\text{energy output}}{\text{energy input}}$

But since the time is the same for both output and input, we can write efficiency as a power ratio.

Efficiency $= \frac{\text{power output}}{\text{power input}}$

$\therefore$ power input $= \frac{\text{power output}}{\text{efficiency}}$

$= \frac{785}{0{\cdot}8}$

$\therefore$ power input $= 981$ W

Exercises

1 A spring has a stiffness of 4 N/mm and is compressed by 100 mm. Find the strain energy stored in the spring.

2 A 40·8 kg mass hanging on a spring has stretched it by 125 mm. Find **i** the strain energy, and **ii** the spring stiffness.

3 Find the potential energy of a 500 kg mass at a height of 60 m.

4 A van carries 25 bags each of which has a mass of 40·8 kg. The floor of the van is 1·5 m above ground level. Find the total potential energy of the bags.

5 An energy input of 10 kJ was required so that a force of 200 N could be exerted over a distance of 40 m. Find **i** the energy output, **ii** the energy lost and **iii** the efficiency.

6 An engine, which was 28% efficient, used 8 litres of fuel. The fuel had a chemical energy of 40 MJ/l. Find the energy output.

7 The kinetic energy of a car travelling along a level road was calculated to be 65 kJ. How far will the car free-wheel along the road against a resistance of 500 N?

8 Find the energy lost in friction when a crate weighing 4 kN is hauled 20 m across a floor. Take $\mu = 0{\cdot}2$.

9 A mass of 40·8 kg is raised to a height of 4 m in 8 s. Find the power required.

10 A force of 750 N moves its point of application a distance of 8 m in 15 s. Find the power.

11 A builder's hoist raises 91·8 kg of materials a height of 15 m in 30 s. Find the power which must be supplied to the hoist motor which is 60% efficient.

12 A van has a mass of 3·06 tonnes and travels against a tractive resistance of 180 N at a speed of 45 km/h on a level road. **i** Find the power developed by the engine. **ii** If the van climbs a slope of 1 in 15 at the same speed, what will the engine power be?

13 The engine of a 1 tonne car develops 2·5 kW at a speed of 90 km/h. Find the tractive resistance.

14 A shaft rotates at 1500 rev/min and transmits a torque of 14 Nm. Find the power.

15 A winch has a drum 300 mm in diameter and can exert a maximum torque of 1·26 kNm at 72 rev/min. Find **i** the maximum power output, **ii** the maximum load (in kg) which the winch could raise vertically, and **iii** the speed at which a maximum load would be raised.

Chapter 8

Mechanical Machines

8.1 A machine

The word 'machine' is used very often in daily life to describe things which do a particular job. We are familiar with machines like adding machines, sewing machines, washing machines, etc. We may think that an engine is a machine, but it depends on what we mean by 'a machine'. In engineering science, a machine is a device which transfers energy. It does not change the form of the energy, it only transfers the energy.

We can consider any machine to be a 'black box': that is a device which takes energy in and puts the energy out in a modified way:

energy input → [machine] → energy output

The symbol for energy is E. Therefore let the symbols for energy input be E_I and for energy output be E_O.

E_I → [machine] → E_O

With mechanical machines, the energy input and the energy output will both be forms of mechanical energy, i.e. work (W). Therefore, for mechanical machines, the black box arrangement is as follows:

work input (W_I) → [machine] → work output (W_O)

From our earlier work, we know that,

mechanical work = force × distance moved,

therefore W_I ($= F_I \times S_I$) → [machine] → W_O ($= F_O \times S_O$)

The work input = input force × distance moved by the input force and the work output = output force × distance moved by the output force.

In many text-books, the input force is called the **effort** and the output force is called the **load**. These alternative names are quite correct and can be used instead.

8.2 Useful ratios

We might ask ourselves why machines are used. The reason, of course, is to enable us to do things we otherwise could not do or, at least, not so easily. A tyre-lever and a crow-bar are examples of the simplest kind of machine – the lever. By using a lever, we can cause a force to be exerted by one end of the lever which is larger than the force we exert on the other end of the lever (remember the Principle of Moments).

We can see from Diagram 8.1, that the effort (F_I) will move much

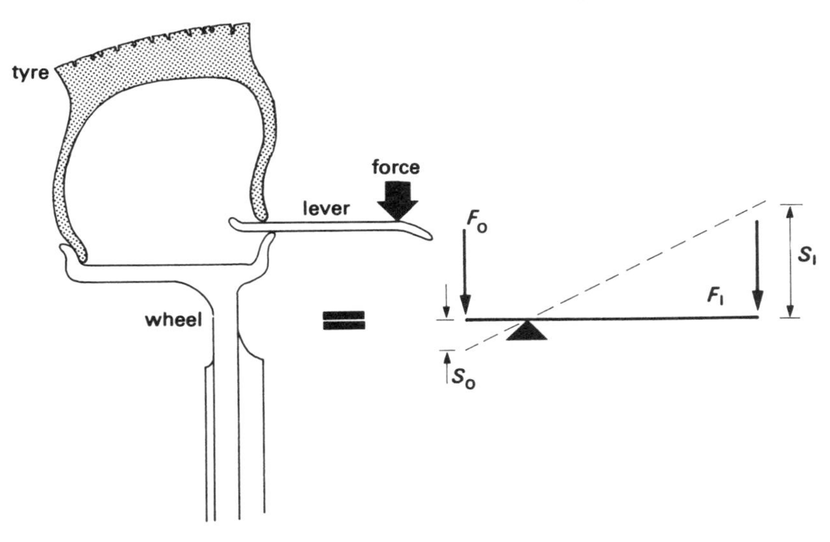

Diagram 8.1

further than the load (F_O) and if we consider the work done at each end of the lever we can see that:

work input = a small effort × a large distance moved
and work output = large load × a small distance moved.

The **force ratio** of load to effort (F_O to F_I) tells us by how much the effort has been multiplied and it is called the **mechanical advantage**.

$$\text{Mechanical advantage} = \frac{\text{load}}{\text{effort}} = \frac{F_O}{F_I}$$

The **distance ratio** of effort distance to load distance (S_I to S_O) tells us how many times further the effort moves compared with the movement of the load. This ratio is commonly called the **velocity ratio**.

$$\text{Velocity ratio} = \frac{\text{distance moved by effort}}{\text{distance moved by load}} = \frac{S_I}{S_O}$$

In the last chapter we dealt with energy inputs and energy outputs and calculated 'how well' the change had been done. We called this the efficiency. The same applies to machines.

$$\text{Efficiency} = \frac{\text{energy output}}{\text{energy input}} \quad \text{(see Chapter 7)}$$
$$= \frac{\text{work output}}{\text{work input}}$$
$$= \frac{\text{load} \times \text{distance moved by load}}{\text{effort} \times \text{distance moved by effort}}$$

In symbols this can be written as:

$$\eta = \frac{E_O}{E_I} = \frac{W_O}{W_I} = \frac{F_O \times S_O}{F_I \times S_I}$$

It is worth noting that most simple mechanical machines are in use as lifting machines – to control and move heavy objects, e.g. there are millions of cars and every one has a car jack.

8.3 Examples of simple machines

i The lever is the simplest machine of all. Next to the lever comes the single fixed pulley as in Diagram 8.2. This kind of arrangement makes the lifting more convenient, but it doesn't make the effort required any smaller, because the fixed pulley acts as a direction changer only, and $S_I = S_O$, therefore velocity ratio = 1.

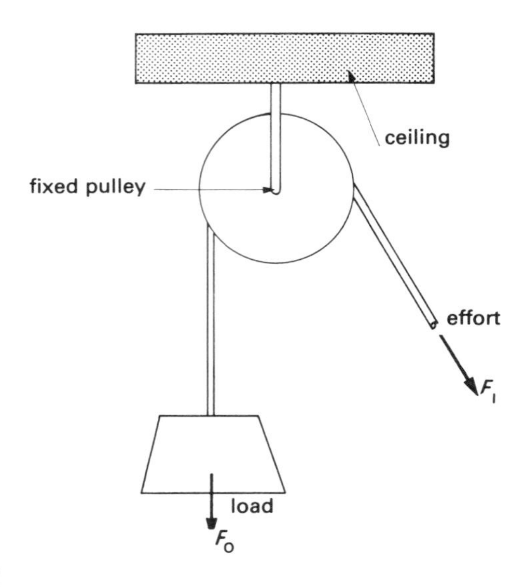

Diagram 8.2

ii The snatch block is a single pulley to which the load is attached. The snatch block moves with the load as shown in Diagram 8.3. The effect of the snatch block is to share the load between two ropes, therefore each rope takes half the load. However, for the load to move say 1

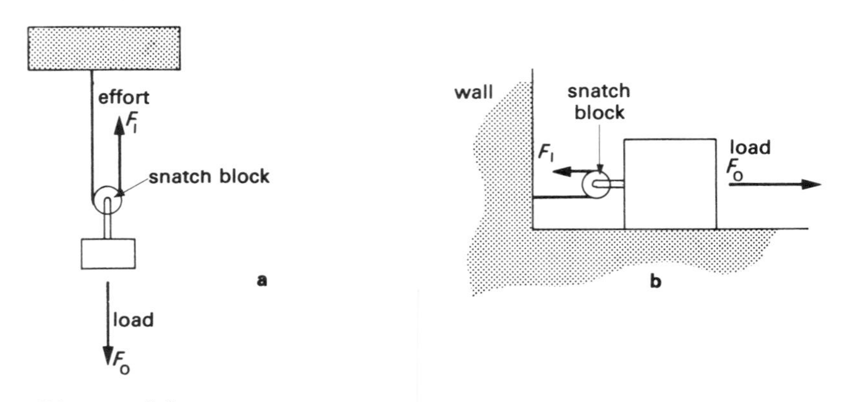

Diagram 8.3

metre then twice as much rope length must be moved and since there is only one free end of rope, the effort end, then the effort must move twice as far as the load.

$$\text{Velocity ratio} = \frac{\text{distance moved by effort}}{\text{distance moved by load}}$$
$$= \frac{2}{1}$$
$$\text{velocity ratio} = 2$$

Sometimes, for convenience, a fixed pulley is used with a snatch block as shown in Diagram 8.4, but this does not change the ratios, it only changes the direction in which the effort acts.

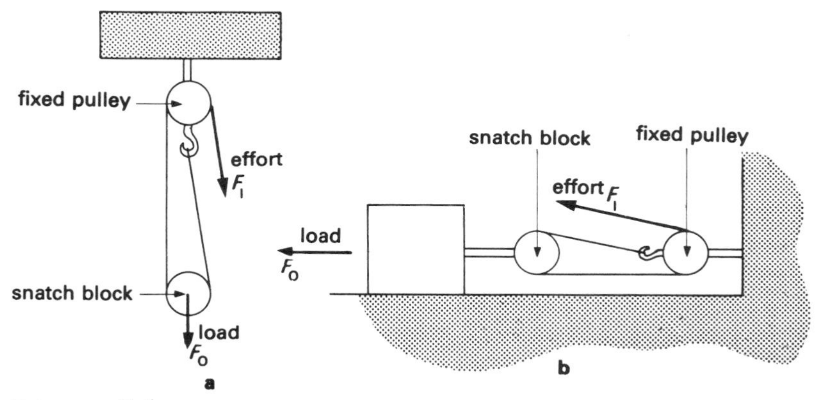

Diagram 8.4

iii The block and tackle is the name given to a machine consisting of several pulleys and a rope. A pulley block is a single unit which may have many pulleys in it (a two-pulley block is shown in Diagram 8.5**a**).

To show clearly on a diagram where the ropes go and how many pulleys there are, we use the method shown in Diagram 8.5**b**. By using smaller circles for successive pulleys in a block we are able to show more clearly the line that the rope follows.

Numerous block and tackle arrangements are possible and some are shown in Diagram 8.6. For each of these arrangements, the number of ropes which share the load (F_O) is the number of ropes cut by the line X–X in the Diagram. Take example **a** in Diagram 8.6.

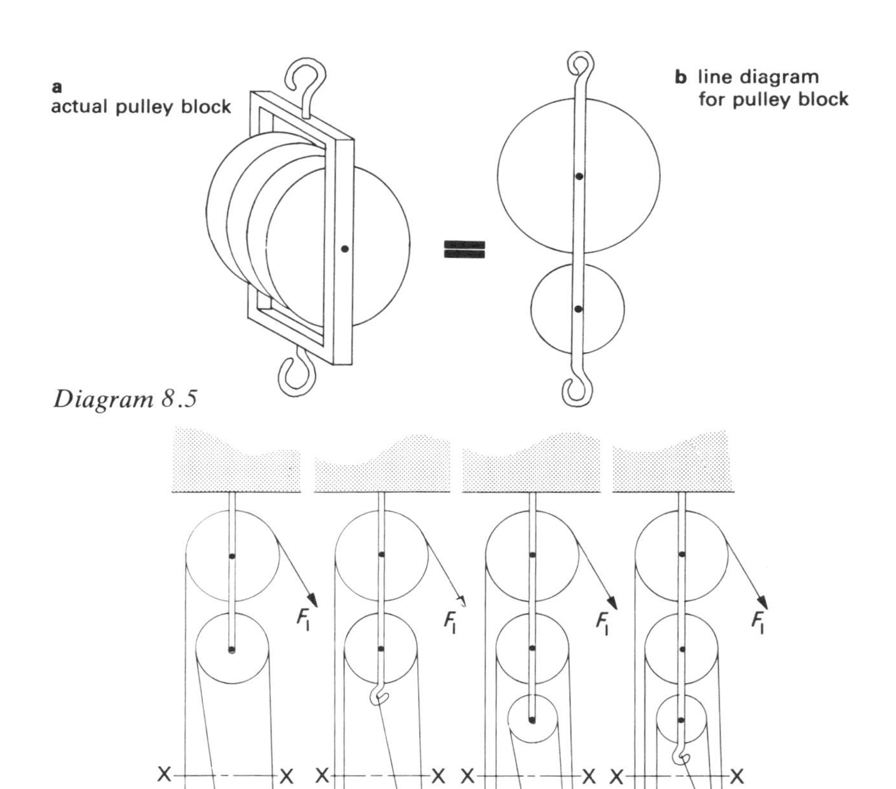

Diagram 8.5

Diagram 8.6

For the load to move 1 metre the three ropes at X–X have each to be moved 1 metre – a total of 3 metres of rope. Since there is only one free end of rope (at the effort) the effort must move 3 metres to make the load move 1 metre.

Velocity ratio = 3

By looking at each of the arrangements we can find their velocity ratios for:

a 3 pulleys; 3 load-sharing ropes: velocity ratio = 3
b 4 pulleys; 4 load-sharing ropes: velocity ratio = 4
c 5 pulleys; 5 load-sharing ropes: velocity ratio = 5
d 6 pulleys; 6 load-sharing ropes: velocity ratio = 6

In general, for a block and tackle, the velocity ratio is always equal to the number of load-sharing ropes. However, be careful if counting the pulleys because this does not **always** give the velocity ratio. Remember the snatch block!

iv Another very simple lifting machine is the winding drum or windlass as often seen at old-fashioned water-wells. A horizontal drum is positioned across the top of the well and it is turned by a handle. The bucket of water is then raised by the rope being wound round the drum (Diagram 8.7).

For the winding drum, velocity ratio $= \dfrac{\text{distance moved by effort}}{\text{distance moved by load}}$

Consider 1 full turn of the drum and handle:

distance moved by effort = circumference of circle traced out by the handle
$= 2\pi R$

and, distance moved by load = amount of rope wound onto the drum
= circumference of drum
$= \pi D$

therefore, velocity ratio $= \dfrac{2\pi R}{\pi D} = \dfrac{2R}{D}$

v The wheel and axle is really very like the winding drum, but instead of an effort handle there is an effort wheel and the drum is now called an axle (Diagram 8.8). Consider 1 revolution of the wheel and axle:

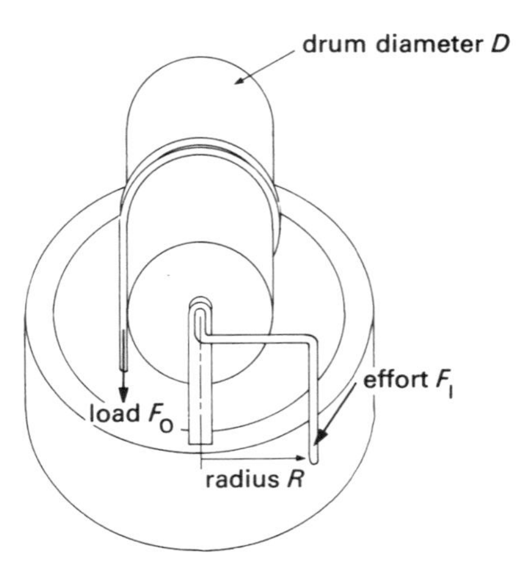

Diagram 8.7

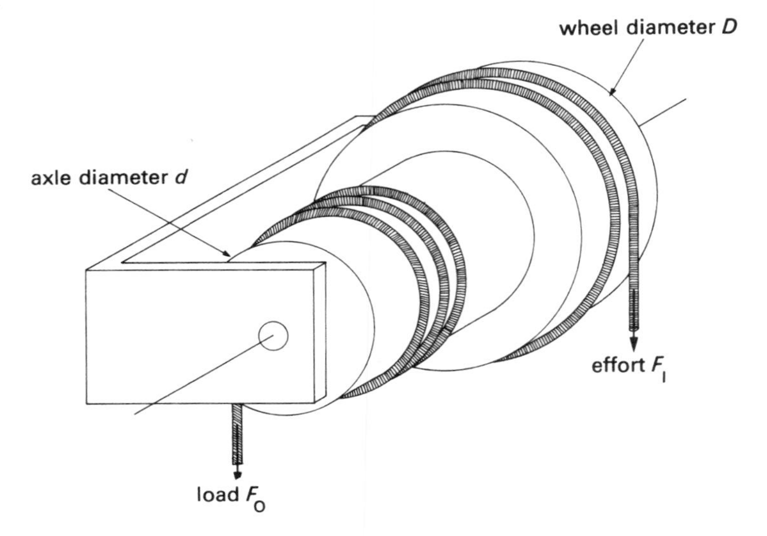

Diagram 8.8

distance moved by effort = circumference of wheel

$= \pi D$

distance moved by load = circumference of axle

$= \pi d$

$$\text{therefore, velocity ratio} = \frac{\text{distance moved by effort}}{\text{distance moved by load}} = \frac{\pi D}{\pi d}$$

$$\therefore \text{velocity ratio} = \frac{D}{d}$$

In each of these simple machines we have been able to work out the velocity ratio because of the physical arrangement of the machines. Whenever we know the size and shape of the machine and how the machine works, we will be able to calculate the velocity ratio (i.e. the distance ratio for effort and load). We could prove this by doing tests on the machines to see if our theory is correct.

8.4 Machine testing

To test any machine, we use a number of different loads and for each one in turn we find the effort required to make the load start to move. Therefore, for each load and its corresponding effort we can find the mechanical advantage.

Also, by measuring the distances moved by the load and the effort we can find the velocity ratio.

A light-weight block and tackle under test gave the following results:

distance moved by each load (S_O) = 1·6 m
distance moved by each effort (S_I) = 400 mm

For the first load (5 N) an effort of 3·5 N was required. Therefore, mechanical advantage at this load is:

$$\text{mechanical advantage} = \frac{\text{load}}{\text{effort}} = \frac{5}{3{\cdot}5}$$

$$\therefore \text{mechanical advantage} = 1{\cdot}43$$

Also, efficiency at this load is:

$$\eta = \frac{\text{load} \times \text{distance moved by load}}{\text{effort} \times \text{distance moved by effort}} = \frac{5 \times 0{\cdot}4}{3{\cdot}5 \times 1{\cdot}6}$$

$$\therefore \eta = 0{\cdot}357 = 35{\cdot}7\%$$

Further loads, each 5 N larger, were used and the values of effort, mechanical advantage and efficiency were tabulated.

Load (F_O) (N)	Effort (F_I) (N)	Force ratio $\left(\frac{F_O}{F_I}\right)$	Efficiency (%)
5	3·5	1·43	35·7
10	4·5	2·22	55·6
15	6	2·5	62·5
20	7	2·86	71·4
25	8	3·13	78·1
30	9·5	3·16	78·9
35	11	3·18	79·5
40	12·5	3·2	80

From the table of results, we can see that as the load gets larger, the effort gets larger too (as we would expect). Also, as the load gets larger, so does the efficiency, but not in equal stages.

These results are best shown on a graph, and the two most useful graphs are **i** effort against load and **ii** efficiency against load. In each case the load is marked along the horizontal axis of the graph (Diagram 8.9**a** and **b**). These graphs show the trends more clearly.

a The effort/load graph shows that the effort increases uniformly with the load and that even with no load, some effort is required just to make the machine work.

b The efficiency/load graph shows that the efficiency increases very rapidly to begin with but gradually flattens off with larger loads. This means that there is a limit to the efficiency of a machine.

A different machine will have a different limit to its efficiency. Some machines will have very high efficiencies (over 80%), some will have medium efficiencies (50 to 80%) and some will have low efficiencies (less than 50%).

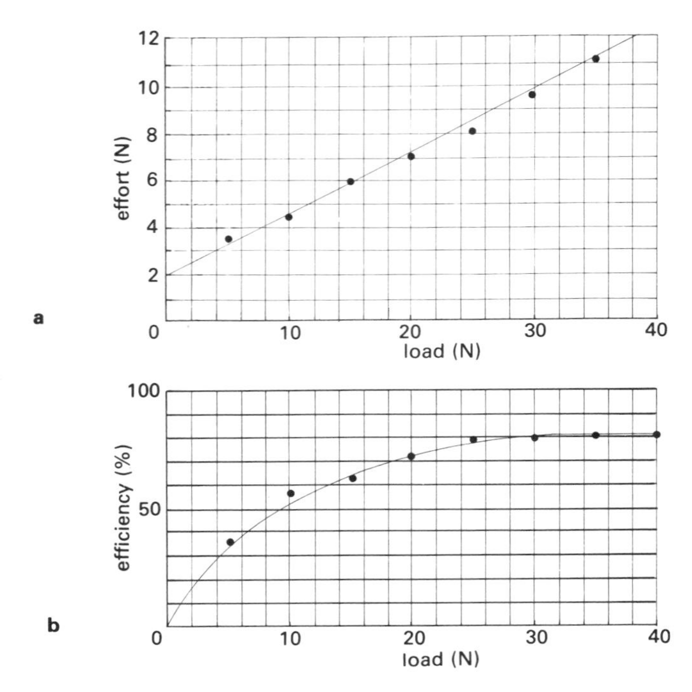

Diagram 8.9

The reason why different machines have different efficiencies is that the amount of friction to be overcome varies from machine to machine. A machine with a very high efficiency has very little friction, but a machine with a low efficiency has quite a lot of friction. A block and tackle for example, is usually very efficient.

Example 1

The input forces and output forces for a machine being tested were as shown below.

input force (F_I) (N)	8	10	12	14	16
output force (F_O) (N)	10	20	30	40	50

Draw the graph of input force to output force and find:

i the input force for an output force of 35 N;

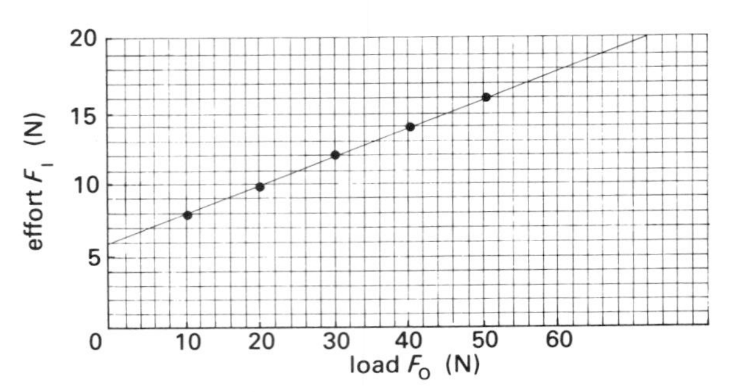

Diagram 8.10

ii the output force for an input force of 9 N;

iii the force ratio when the output force is 30 N;

iv the input force for no output force.

Solution

From the graph (Diagram 8.10):

i when the output force = 35 N
the input force = 13 N

ii when the input force = 9 N
the output force = 15 N

iii force ratio $= \dfrac{\text{output force}}{\text{input force}}$

when output force = 30 N, input force = 12 N

$\therefore$ force ratio $= \dfrac{30}{12}$
$= 2{\cdot}5$

iv when output force = 0 (i.e. at the origin)
the input force = 6 N

Example 2

The graph of effort/load and efficiency/load for a machine under test are shown in Diagram 8.11.

i Find the effort to overcome the friction in the machine.

ii When the load is 300 N, find the effort.

iii When the effort is 13 N, find the load.

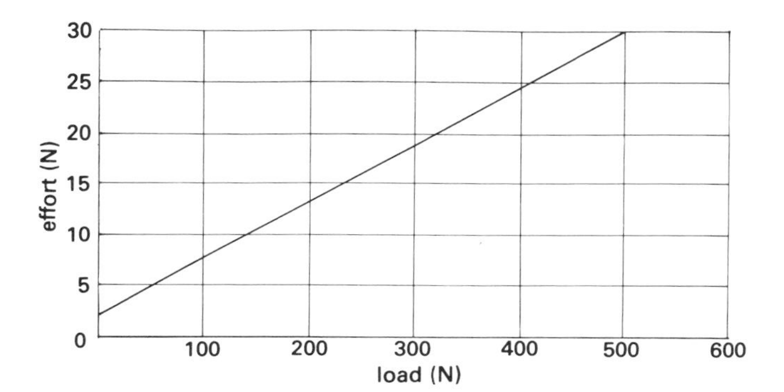

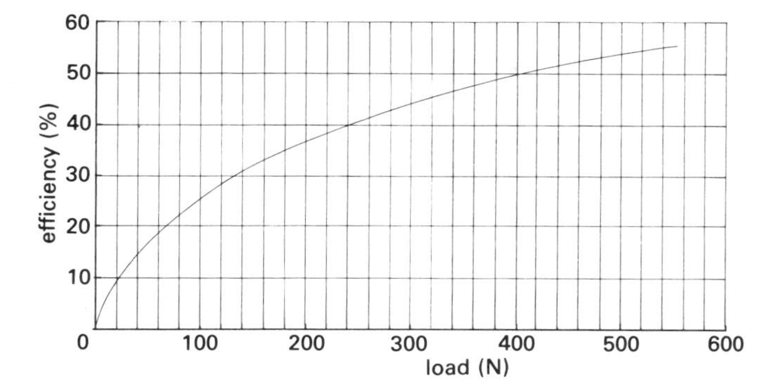

Diagram 8.11

iv When the load is 400 N, find the efficiency.
v When the efficiency is 40%, find the mechanical advantage.

Solution

i Effort (F_I) to overcome friction $= 2$ N
ii When $F_O = 300$ N, $F_I = 18{\cdot}5$ N
iii When $F_I = 13$ N, $F_O = 200$ N
iv When $F_O = 400$ N, $\eta = 50\%$
v When $\eta = 40\%$, $F_O = 240$ N
and when $F_O = 240$ N, $F_I = 15$ N

$$\text{therefore mechanical advantage} = \frac{F_O}{F_I} = \frac{240}{15}$$

$$\therefore \text{mechanical advantage} = 16$$

8.5 Rotating machines

In our work on machines we have considered only straight line motion for effort and load, but 'simple machines' also include machines which transmit torque. From Chapter 6, we know that the work done by a torque is given by:

work done = torque × number of revolutions × 2π
i.e. $W = 2\pi NT$

Therefore, considering a rotating machine as a black box:

input W_I → machine → W_O output
$(= 2\pi N_I T_I)$ $(= 2\pi N_O T_O)$

In this case, instead of using force ratio and distance ratio, it is more appropriate to use torque ratio and speed ratio:

$$\text{torque ratio} = \frac{\text{torque output}}{\text{torque input}} = \frac{T_O}{T_I}$$

$$\text{and speed ratio} = \frac{\text{speed input}}{\text{speed output}}$$

$$= \frac{\text{number of input revolutions}}{\text{number of output revolutions}} \text{ in the same time}$$

$$\text{speed ratio} = \frac{N_I}{N_O}$$

As in all machines, we will be concerned with the efficiency of the transfer of energy. With rotating machines, it is more common to deal with rotational speeds in rev/min or rev/sec (n) rather than the total number of revolutions (N).

From Chapter 7, the power transmitted by a torque is given by:

$P = 2\pi nT$

Therefore, for rotating machines, there are a number of ways of finding the efficiency:

$$\text{efficiency } (\eta) = \frac{\text{energy output}}{\text{energy input}} = \frac{\text{work output}}{\text{work input}} = \frac{\text{power output}}{\text{power input}}$$

$$\therefore \eta = \frac{2\pi N_O T_O}{2\pi N_I T_I}$$

Example 3

A gearbox for a crane was operating under the following conditions:

input torque = 60 N m
input speed = 3000 rev/min
output torque = 240 N m
output speed = 563 rev/min

Find the operating efficiency of the gearbox.

Solution

$$\text{Efficiency} = \frac{\text{power output}}{\text{power input}} = \frac{2\pi N_O T_O}{2\pi N_I T_I} = \frac{2\pi \times 563 \times 240}{2\pi \times 3000 \times 60}$$

$$\therefore \eta = 0{\cdot}75$$

Example 4

A compressor has to run at 540 rev/min and it is to be driven by an engine which runs at 1500 rev/min when delivering a torque of 30 N m. The gearbox which is used has an efficiency of 90% under these conditions. Find the torque supplied to the compressor.

Solution

$$\text{Efficiency} = \frac{\text{power supplied to compressor}}{\text{power supplied by engine}}$$

$$\therefore \eta = \frac{2\pi n_c T_c}{2\pi n_e T_e}$$

$$\therefore \frac{90}{100} = \frac{540 \times T_c}{1500 \times 30}$$

$$\therefore 0{\cdot}9 = 0{\cdot}012 \times T_c$$

$$\therefore T_c = \frac{0{\cdot}9}{0{\cdot}012}$$

$\therefore$ compressor torque (T_c) = 75 N m

8.6 Gears

The gears which you are probably most familiar with are bicycle gears. The toothed gear wheels on a bicycle are called sprockets; the crank sprocket is the large one at the pedals and the wheel sprocket is the small one on the rear wheel hub. These two sprockets, connected by the drive chain, have a different number of teeth on them. The space between teeth on each wheel is the same because the links of the chain must fit smoothly on both sprockets.

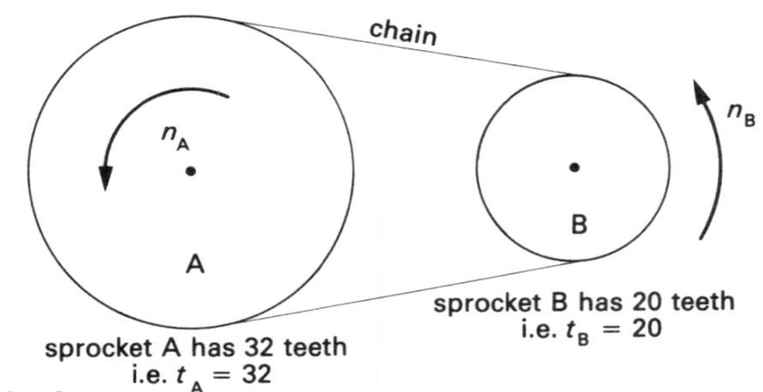

Diagram 8.12

Consider a chain drive as shown in Diagram 8.12. We can see from the diagram that both sprockets will turn in the same direction but not at the same speed. Let the rotational speed of A be n_A rev/s and let the rotational speed of B be n_B rev/s. Then the speed ratio of A to B is given by:

$$\text{speed ratio} = \frac{n_A}{n_B}$$

If we know the number of teeth on A and B then we can calculate the speed ratio. Let the sprocket A make 1 complete revolution. Therefore it will move 32 chain links (the same number of links as teeth on A). Therefore 32 chain links must also have moved sprocket B, but B has only 20 teeth and so B must make 1 revolution (20/20) plus 12/20 of another revolution.

$$\therefore 1 \text{ revolution of A} = \frac{32}{20} \text{ revolutions of B}$$

$$\text{or } 20 \text{ revolutions of A} = 32 \text{ revolutions of B}$$

$$\therefore \frac{\text{speed of A}}{\text{speed of B}} = \frac{20}{32} = \frac{\text{teeth on B}}{\text{teeth on A}}$$

$$\therefore \text{speed ratio (of A to B)} = \frac{n_A}{n_B} = \frac{t_B}{t_A}$$

Example 5

The crank sprocket on a bicycle has 48 teeth and it turns at 60 rev/min. The wheel sprocket has 24 teeth. Find the rotational speed of the wheel.

Solution

$$\text{Speed ratio} = \frac{\text{speed of wheel sprocket}}{\text{speed of crank sprocket}} = \frac{\text{teeth on crank sprocket}}{\text{teeth on wheel sprocket}}$$

$$\therefore \frac{n_w}{n_c} = \frac{t_c}{t_w}$$

where $n_c = 60$ rev/min

$t_c = 48$

and $t_w = 24$

$$\therefore \frac{n_w}{60} = \frac{48}{24}$$

$\therefore$ speed of wheel $(n_w) = 120$ rev/min

Spur gears

When two gear wheels mesh directly with each other, the teeth have to be a special shape. Gear wheels like these are called **spur gears**. Two such gear wheels in mesh are shown in Diagram 8.13. The first thing to notice is that gear wheels A and B will rotate in opposite directions: A clockwise and B anticlockwise in Diagram 8.13.

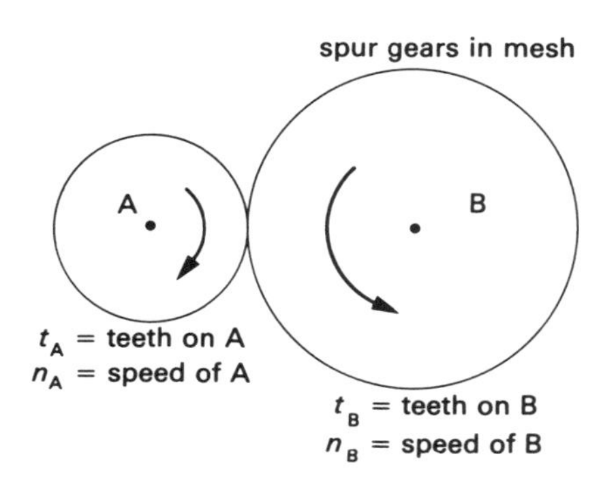

Diagram 8.13

Spur gear wheels in mesh

$$\text{Again, speed ratio (of A to B)} = \frac{n_A}{n_B} = \frac{t_B}{t_A}$$

An important feature of sprockets and spur gears is that all the teeth are whole teeth. We could not have, for example, a gear with 42.9 teeth.

Example 6

An electric motor which runs at 1200 rev/min is used to drive a winch which has to run at 900 rev/min. A 30 tooth gear wheel is to be used with one other gear wheel. Find **i** the tooth size for the other gear wheel, and **ii** which gear is to be connected to the motor shaft and which to the winch shaft.

Solution

i Speed ratio $= \dfrac{\text{speed of motor}}{\text{speed of winch}} = \dfrac{\text{teeth on winch gear}}{\text{teeth on motor gear}}$

$$\therefore \frac{1200}{900} = \frac{t_w}{t_m}$$

$$\therefore \frac{t_w}{t_m} = \frac{4}{3}$$

One gear wheel has 30 teeth, therefore, for a tooth ratio of 4 to 3, the other gear wheel must have either 40 teeth or $22\frac{1}{2}$ teeth. Since gear wheels cannot have half teeth, we must use a 40 tooth gear wheel. Therefore the other gear wheel must have 40 teeth.

ii To have the correct speed ratio,

$$\frac{t_w}{t_m} = \frac{4}{3} = \frac{40}{30}$$

Therefore the winch has the 40 tooth gear and the motor has the 30 tooth gear.

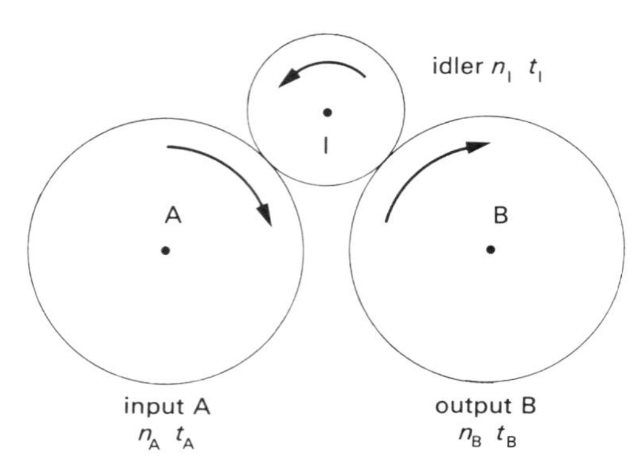

Diagram 8.14

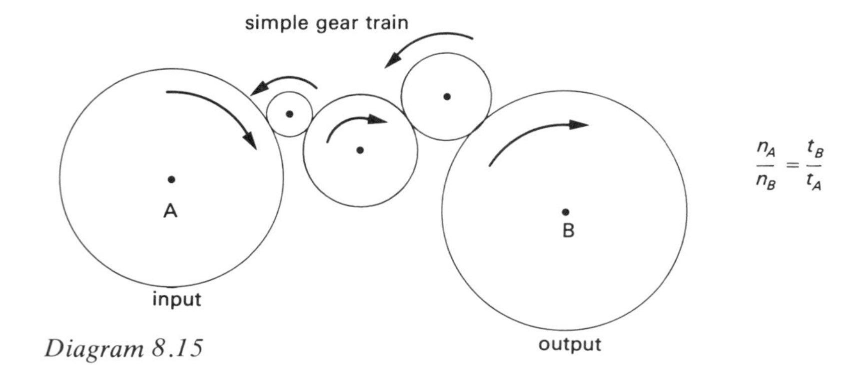

Diagram 8.15

Simple gear trains and idler gears

Whenever it is important to have the input rotation in the **same** direction as the output rotation, we will have to use a small intermediate gear called an idler gear. This does not affect the input/output speed ratio, it only changes the **direction** of the rotation of the output (rather like reverse gear in a car gearbox). A possible arrangement is shown in Diagram 8.14. Consider the speed ratio of gear A to the idler gear:

speed ratio $= \dfrac{n_A}{n_I} = \dfrac{t_I}{t_A}$

speed of idler gear, $n_I = n_A \times \dfrac{t_A}{t_I}$

Now consider the speed ratio of the idler gear to gear B:

speed ratio $= \dfrac{n_I}{n_B} = \dfrac{t_B}{t_I}$

speed of idler gear, $n_I = n_B \times \dfrac{t_B}{t_I}$

We now have two expressions for the same thing, n_I, and they must obviously be equal.

$$\therefore n_A \times \frac{t_A}{t_I} = n_B \times \frac{t_B}{t_I}$$

$$\therefore n_A\, t_A = n_B\, t_B$$

$$\text{or } \frac{n_A}{n_B} = \frac{t_B}{t_A}$$

Therefore, for a series of gears in mesh, the speed ratio is given by the ratio of the first gear to the last gear.

$$\text{Speed ratio} = \frac{n_A}{n_B} = \frac{t_B}{t_A}$$

(The idler gear has no effect on the ratio.) In fact this is true no matter how many idler gears we use, one after the other, in between the input and output (see Diagram 8.15).

Compound gear train

When we need to have a quite large speed ratio then using a simple gear train would take up a lot of space. It would also mean that we have to use a very large gear wheel. For example, if we required a speed ratio of 40 to 1 then the large gear would need to have 40 times as many teeth as the small gear. Since the smallest normal spur gear has 18

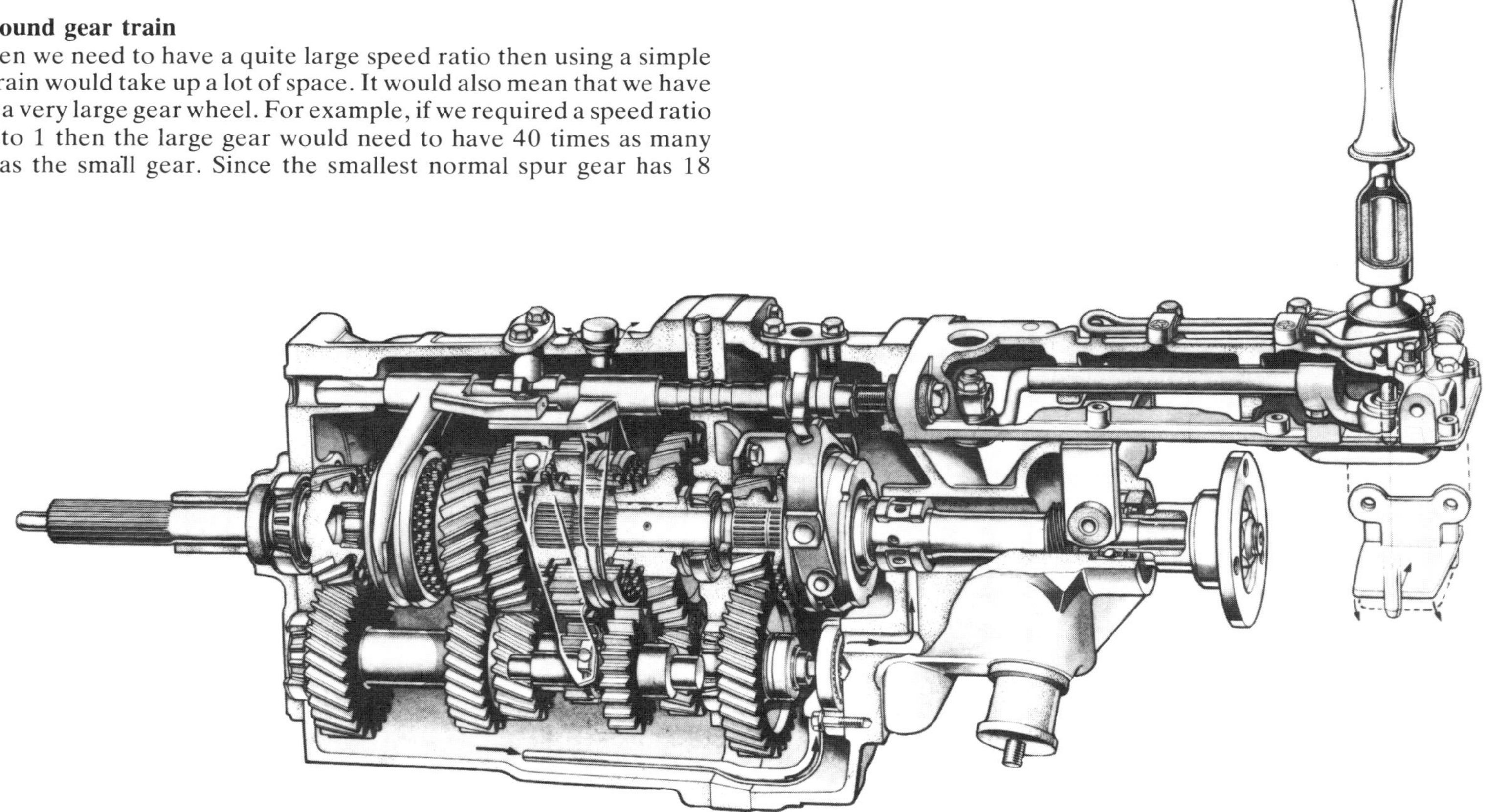

Section of a Jaguar, Rover, Triumph five-speed gear box

teeth, then the large gear in this instance would need to have 720 teeth. This is very large, and could be over 1 m in diameter, even with very fine teeth.

To solve this kind of problem, we can use what is called a compound gear train. This is an arrangement of separate shafts, at least one of which has two different gears on it. Diagram 8.16 shows a typical arrangement.

The input shaft turns gear A which drives gear B. Because gear B and C are on the same lay shaft they must both turn at the same time and at the same speed. Then gear C drives gear D which gives us the output:

input → A → lay shaft → D → output
B/C

As the drive is transferred from shaft to shaft the direction of rotation will change (as for a simple gear train). Therefore in Diagram 8.16, the input and output shafts will turn in the same direction and the lay shaft will turn in the opposite direction. To find the overall speed ratio of the input to the output, we can say:

overall speed ratio = speed ratio of input to lay shaft × speed ratio of lay shaft to output

$$\frac{n_A}{n_D} = \frac{n_A}{n_{BC}} \times \frac{n_{BC}}{n_D}$$

Or, remembering that speed ratios can be calculated if we know the sizes of the gear wheels:

$$\frac{n_A}{n_D} = \frac{t_B}{t_A} \times \frac{t_D}{t_C}$$

Example 7

Find the overall speed ratio of input to output for the compound gear train shown in Diagram 8.17.

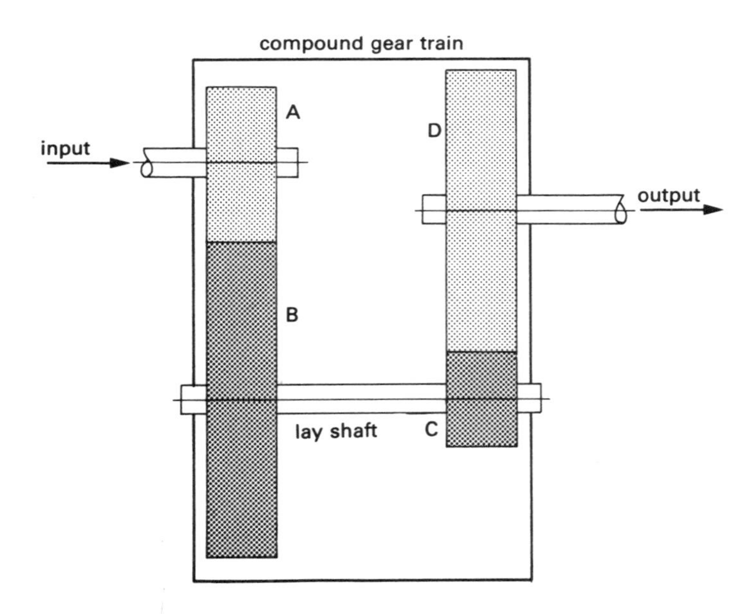

Diagram 8.16

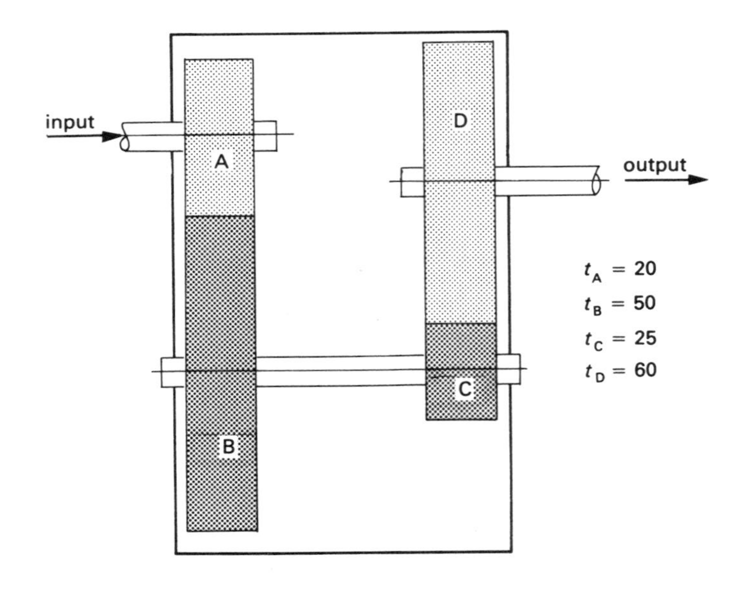

Diagram 8.17

Solution

$$\text{Overall speed ratio} = \frac{\text{speed of A}}{\text{speed of B}} \times \frac{\text{speed of C}}{\text{speed of D}}$$
$$= \frac{t_B}{t_A} \times \frac{t_D}{t_C}$$
$$= \frac{50}{20} \times \frac{60}{25}$$
$$\therefore \text{overall speed ratio} = 6$$

Example 8

A motor which runs at 1500 rev/min drives a pump at 400 rev/min. The compound gear train which is used is shown in Diagram 8.18. Find the number of teeth which must be on gear wheel C.

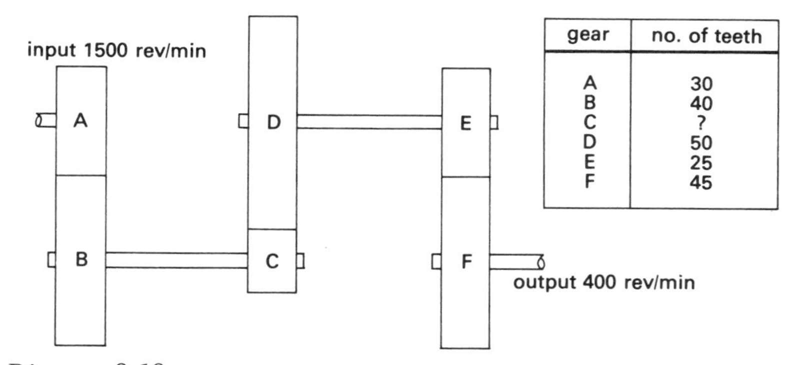

gear	no. of teeth
A	30
B	40
C	?
D	50
E	25
F	45

Diagram 8.18

Solution

$$\text{Overall speed ratio} = \frac{n_A}{n_B} \times \frac{n_C}{n_D} \times \frac{n_E}{n_F}$$
$$\therefore \frac{n_A}{n_F} = \frac{t_B}{t_A} \times \frac{t_D}{t_C} \times \frac{t_F}{t_E}$$
$$\therefore \frac{1500}{400} = \frac{40}{30} \times \frac{50}{t_C} \times \frac{45}{25}$$
$$\therefore 3{\cdot}75 = \frac{120}{t_C}$$
$$\therefore t_C = \frac{120}{3{\cdot}75}$$
$$= 32$$

∴ gear wheel C must have 32 teeth

8.6 Screw-threads

A screw-thread is simply an inclined plane wrapped round a rod or cylinder. In the same way, a helter-skelter is simply a straight chute wrapped round a vertical column.

The most common use of screw-threads as machines are, for example, a car jack and a vice. One complete turn of the screw makes the nut move one thread further along the screw.

The distance along the axis from one thread to the next is called the pitch (p) and is usually measured in mm. Machines which use screw-threads are treated in exactly the same way as other simple machines:

$$\text{velocity ratio} = \frac{\text{distance moved by effort}}{\text{distance moved by load}}$$

Consider 1 revolution of the effort handle of the screw-jack shown in Diagram 8.19.

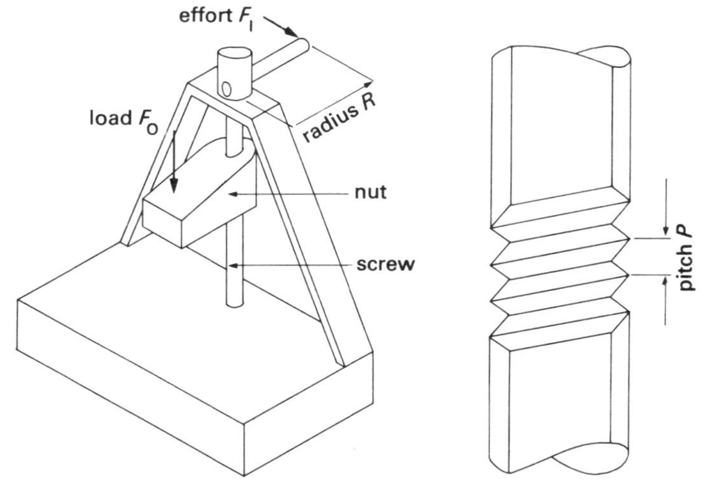

Diagram 8.19

Distance moved by effort $= 2\pi R$
and distance moved by load = pitch (p)

$\therefore$ velocity ratio $= \dfrac{2\pi R}{p}$

Example 9

The following information was noted while carrying out a test on a screw-jack: radius of effort handle = 210 mm; pitch of screw-thread = 5 mm; effort used (F_I) = 19 N; load raised (F_O) = 1760 N. Find:

i the velocity ratio;
ii the mechanical advantage at this load;
iii the work input for 10 revolutions of the effort handle;
iv the work output for 10 revolutions of the effort handle;
v the efficiency at this load.

Solution

i To find velocity ratio, consider 1 revolution of effort handle:

distance moved by effort $= 2\pi \times 210 = 1320$ mm
and distance moved by load = pitch (p) $\times$ 5 mm

$\therefore$ velocity ratio $= \dfrac{1320}{5} = 264$

ii Mechanical advantage $= \dfrac{\text{load}}{\text{effort}} = \dfrac{F_O}{F_I} = \dfrac{1760}{19}$

$\therefore$ mechanical advantage = 92·6

iii For 10 revolutions of effort handle:
work input = force (F_I) $\times$ distance moved
$= 19 \times (10 \times 2\pi \times 210 \times 10^{-3})$
$\therefore W_I = 251$ J

iv For 10 revolutions of effort handle, screw turns 10 times:
$\therefore$ work output = force (F_O) $\times$ distance moved
$= 1760 \times (10 \times 5 \times 10^{-3})$
$W_O = 88$ J

v Efficiency $= \dfrac{\text{work output}}{\text{work input}} = \dfrac{88}{251}$

$\therefore \eta = 0{\cdot}351$ or 35·1%

Example 10

The cutting tool on a lathe is moved horizontally by a screw which rotates at 300 rev/min. The pitch of the screw is 2 mm and the force required at the cutting tool is 500 N (Diagram 8.20). Under these conditions, the screw has an efficiency of 40%. Find:

i the linear speed of the tool (in mm/s);
ii the power output (at the tool);
iii the power input (to the screw).

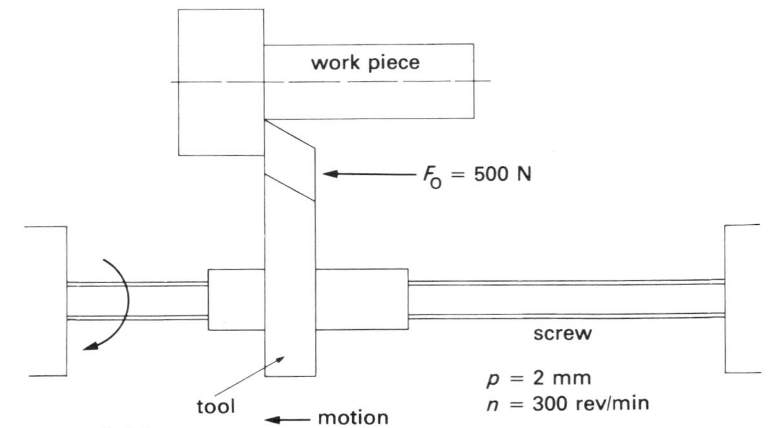

Diagram 8.20

Solution

i In 1 second, the screw turns $\dfrac{300}{60}$ times, i.e. 5 times

$\therefore$ in 1 second, distance moved by tool $= 5 \times$ pitch
$= 5 \times 2$
$= 10$ mm

$\therefore$ linear speed of tool = 10 mm/s

ii Power output

$$(P_O) = \frac{\text{force} \times \text{distance moved}}{\text{time taken}}$$
$$= \text{force} \times \text{speed}$$
$$= 500 \times (10 \times 10^{-3})$$
$$\therefore P_O = 5 \text{ W}$$

iii Power input

$$(P_I) = \frac{\text{power output}}{\text{efficiency}}$$
$$= \frac{5}{0{\cdot}4}$$
$$\therefore P_I = 12{\cdot}5 \text{ W}$$

8.7 Overhauling (running back)

Very often, with simple lifting machines, the effort cannot be taken away otherwise the load will fall back. This is particularly true for a block and tackle or a wheel and axle. What happens is that the load makes the machine run in the reverse direction. It can only do this if there is not too much friction in the machine.

As a rough guide, we can say that a machine with an efficiency greater than 50% is likely to run back (or overhaul). A machine with an efficiency less than 50% is not likely to run back (or overhaul). Some lifting machines with an efficiency greater than 50% have a non-return mechanism or ratchet or some other device to prevent the machine running back and lowering the load unexpectedly.

Exercises

1 A mass of 51 kg was raised by an effort of 25 N. Find the mechanical advantage.

2 When using a machine, the load moved 50 mm while the effort moved 275 mm. Find the velocity ratio.

3 The work output from a machine was 260 J and the work input was 400 J. Find the efficiency.

4 The efficiency of a machine was 45% when moving a load of 800 N. If the load was moved a distance of 0·5 m find the necessary work input.

5 During a test on a machine the following details were noted:
input force = 15 N; input distance moved = 5 m;
output force = 400 N; output distance moved = 150 mm.
Find: **i** the mechanical advantage; **ii** the velocity ratio; **iii** the work input; **iv** the work output; **v** the efficiency.

6 A machine was used to raise a load of 2 kN a distance of 3 m. The effort required was 125 N and the machine efficiency at this load was 45%. Find: **i** the mechanical advantage; **ii** the distance moved by the effort; **iii** the velocity ratio.

7 A rope and a single pulley are used to raise a load of 210 N. The single pulley is arranged to be a snatch block and the effort required is 107 N. Find: **i** the velocity ratio; **ii** the efficiency, if the load moves 5 m.

8 Draw a line diagram to represent a block and tackle with a velocity ratio of 4. Such a block and tackle raises a load of 420 N for an effort of 120 N. Find the efficiency of this machine.

9 In a simple wheel and axle, the wheel has a radius of 125 mm and the axle has a radius of 25 mm. The load was 80 N and the efficiency was 85%. Find the effort.

10 A wheel and axle is to be designed to raise a load of 40 N for an effort of 8 N. The axle is to have a radius of 30 mm and it is estimated that the efficiency will be 90%. Find the necessary radius of the wheel.

11 A windlass has an effort handle at a radius of 300 mm and a drum diameter of 200 mm. It takes an effort of 25 N to raise a mass of 5 kg. The mass is raised a total distance of 3·18 m. Find: **i** the distance moved by the effort handle and **ii** the efficiency.

12 A machine has a velocity ratio of 10. Under test the following results were obtained:

Effort (F_I) in N	28	38	48	58	68
Load (F_O) in N	100	200	300	400	500

Draw the graph of effort/load. Find: **i** the effort required to overcome friction in the machine and **ii** the efficiency at a load of 250 N.

13 Diagram 8.21 shows a combined graph of effort/load and efficiency/load. From the graphs find: **i** the effort required at 0 load; **ii** the effort at a load of 50 N; **iii** the efficiency at this load.

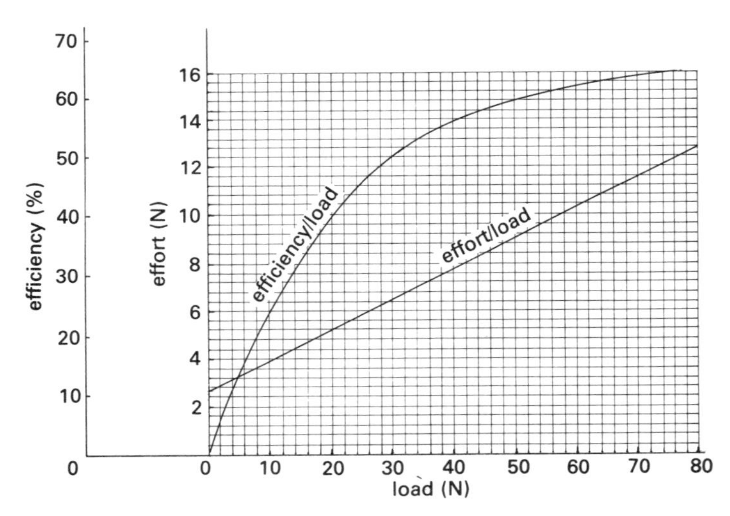

Diagram 8.21

14 Find the efficiency of a gearbox which transfers an input torque of 20 N m on a shaft running at 400 rev/min to an output torque of 70 N m on a shaft running at 80 rev/min.

15 A machine gear box has an efficiency of 85%. The input torque is 14 N m and the input shaft speed is 1500 rev/min. Find: **i** the power input; **ii** the power output; **iii** the output torque for an output shaft speed of 200 rev/min.

16 The crank sprocket on a bicycle has 48 teeth and the wheel sprocket has 16 teeth. The pedals are at a radius of 150 mm and the average total pedal force is 200 N. The bicycle wheels are 660 mm in diameter, and as a machine, the bicycle is 90% efficient. The cyclist turns the pedals at 75 rev/min. Find: **i** the rotational speed of the wheels; **ii** the output driving force at the rim of the wheel; **iii** the linear speed of the bicycle (in km/h).

17 The speed ratio of gears A to C in the simple gear train shown in Diagram 8.22 is 0·4. Find the number of teeth on gear C.

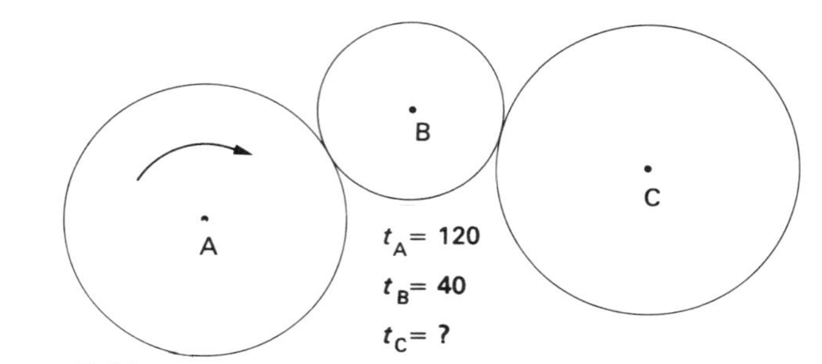

Diagram 8.22

18 For the simple gear train shown in Diagram 8.23, find: **i** the gear which rotates in the same direction as A; **ii** the speed ratios of A to B, A to C and A to D; **iii** the speed of D, if A rotates at 1400 rev/min.

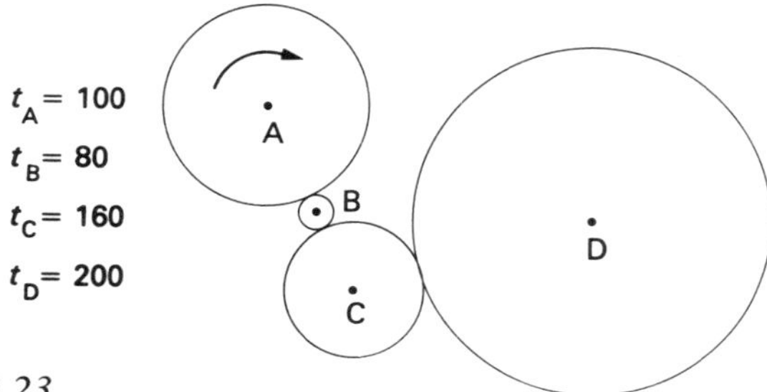

Diagram 8.23

19 The compound gear train shown in Diagram 8.24 is driven by a motor which runs at 750 rev/min. Find **i** the speed ratio of the motor spindle speed to the output shaft speed, and **ii** the output shaft speed.

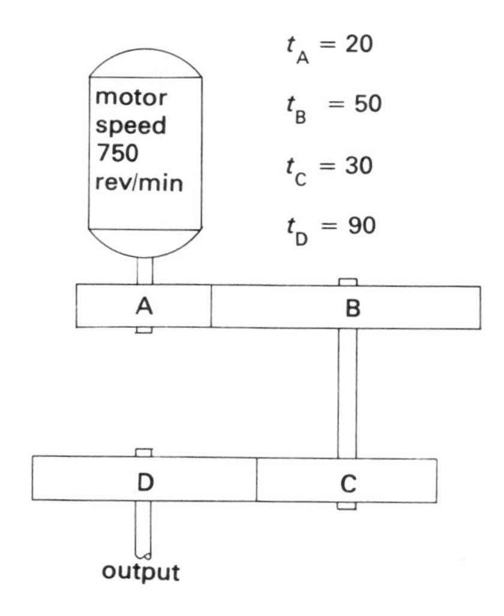

Diagram 8.24

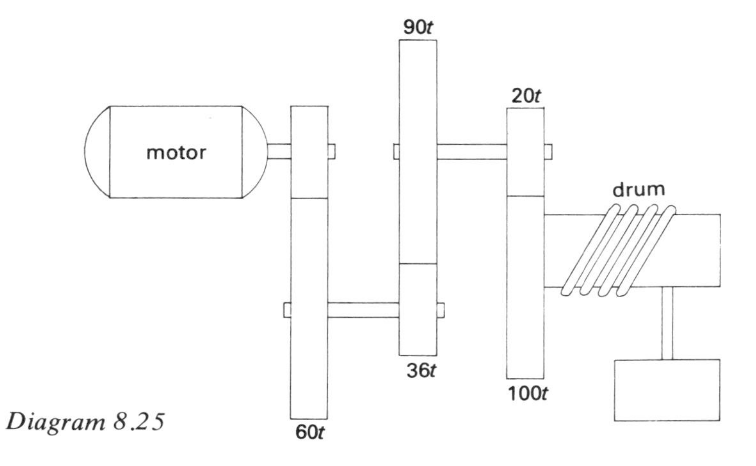

Diagram 8.25

20 A motorised winch is shown in Diagram 8.25. The motor runs at 1350 rev/min and the drum is to rotate at 54 rev/min. Find the number of teeth which the motor drive gear must have to satisfy this requirement.

21 The lifting drum of the motorised winch shown in Diagram 8.25 has a diameter of 200 mm. The motor supplies a torque of 6 N m and the winch has an overall efficiency of 75%. (Other details as in question **20**.) Find: **i** the hoisting speed of the drum (in m/s); **ii** the power input; **iii** the power output; **iv** the load being raised.

22 The efficiency of the crab-winch shown in Diagram 8.26 is 84% at a load of 51 kg. Find **i** the necessary effort at the handle and **ii** the hoisting speed, if the handle is turned once every second.

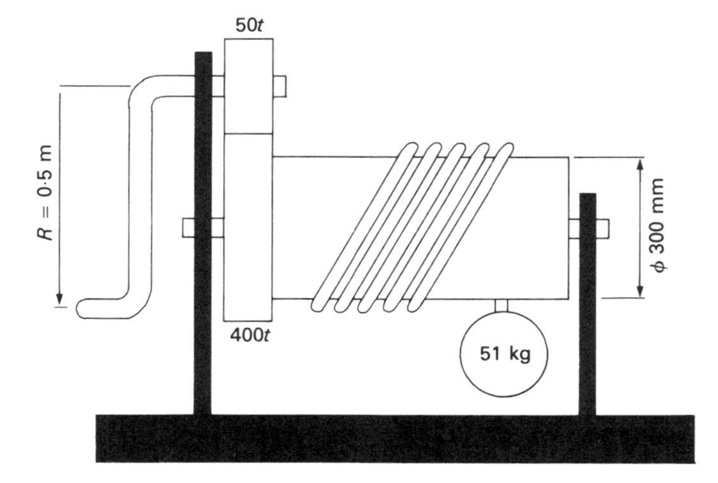

Diagram 8.26

23 The double-purchase crab winch shown in Diagram 8.27 raises a load a distance of 4·81 metres for 60 full turns of the effort handle. Find **i** the overall velocity ratio, and **ii** the number of teeth on the lifting drum gear.

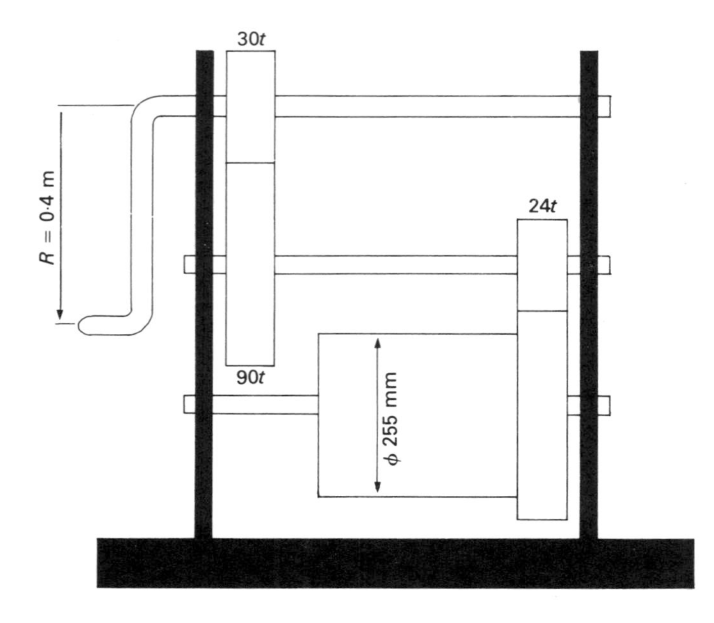

Diagram 8.27

24 A screw-jack for a car is designed to raise a maximum load of 5 kN. The screw-thread has a pitch of 5 mm, the radius of the effort handle is 250 mm and the efficiency at maximum load is 28%. Find: **i** the velocity ratio and **ii** the effort required to raise the maximum load.

25 A vertical screw driven hoist raises a load of 102 kg when the screw rotates at 1200 rev/min. The pitch of the screw is 4 mm and the input torque to the screw is 2·4 N m. Find: **i** the speed at which the load is raised; **ii** the power output of the hoist; **iii** the power input to the hoist; **iv** the efficiency of the hoist at this load; **v** how long it will take to raise the load 5 m.

Chapter 9

Electricity

9.1 Electricity

All matter is made up of atoms and each atom is made up of a nucleus with electrons orbiting around it. Electrons are considered to be negatively charged particles. Some materials have electrons which can be moved quite easily. When an electron is moved from one atom to the next, it pushes another electron to the next again atom and so on. This is rather like having a tube filled with marbles and then pushing another marble into one end of the tube. All the marbles in the tube move together and one will fall out of the other end of the tube just as the first marble is being pushed in (see Diagram 9.1). The marbles can be considered to behave just like 'loose' electrons (or 'free' electrons as they are called).

Electrons are so small, that to measure their flow would be an almost impossible task. The quantity of electrons which are being moved is called the **charge** (symbol Q) and this is measured in units called coulombs (symbol C).

1 unit of charge = 1 coulomb = 6289×10^{15} electrons

Charge, therefore, is the amount or quantity of electrons. How quickly the electrons are made to flow is the rate of flow of charge. This is called the **electric current**:

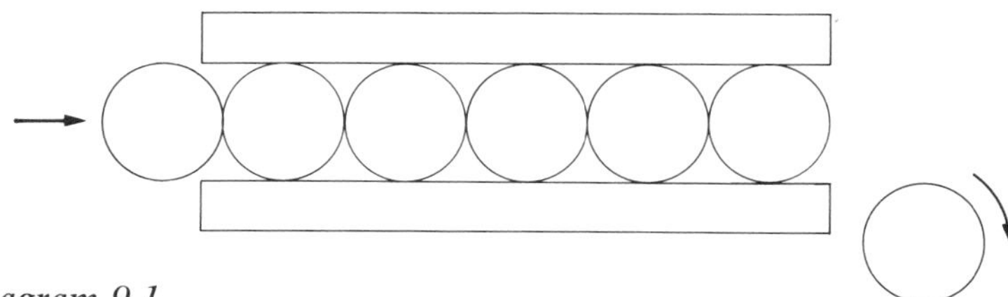

Diagram 9.1

$$\text{current} = \frac{\text{quantity of electrons}}{\text{time}} = \frac{\text{charge (in coulombs)}}{\text{time (in seconds)}}$$

Therefore electric current (symbol I) is measured in coulombs per second, and this unit is given a special name – the ampere (symbol A) which is sometimes shortened to amp. Therefore, 1 C/s = 1 A.

Because 1 amp is quite a large current, we will quite often use submultiples such as milliamp (mA) and microamp (μA).

So far, then, we have looked at charge and electric current. The relationship between these two quantities is usually written as:

charge, Q = current (I) $\times$ time (t)
or $\quad Q = It$

Example 1

A charge of 50 C is transferred in 10 s. Find the electric current.

Solution

$$\text{Current} = \frac{\text{charge}}{\text{time}} = \frac{50}{10} = 5 \text{ C/s or A}$$

$$\therefore I = 5 \text{ A}$$

Example 2

A current of 65 mA was passed through a circuit for 10 min. Find the charge.

Solution

$$Q = It = (65 \times 10^{-3}) \times (10 \times 60)$$

$$\therefore Q = 39 \text{ C}$$

In practical circuits we use an instrument called an ammeter to measure the rate of flow of charge, i.e. to measure the electric current. There are a great number of different types of meter available, so we must be sure that the meter we are using is suitable, i.e. that it will read what we want it to read.

You will discover, in your practical work, that a very sensitive ammeter can be used for different purposes in a wide range of situations.

9.2 Effects of an electric current

We are familiar with what electricity does for us in the home. It gives us light at night, heat from electric fires, hot water from immersion heaters. It powers electric motors which drive washing machines, food mixers, electric fans, sewing machines, electric trains and other devices. There are many more examples of how electricity is used both in the home and in industry, but they can all be grouped under three main effects of an electric current.

a Heating and lighting

These are really the same effect. As current flows through the bar of an electric fire it makes the bar heat up, usually to an orange/red light. In light bulbs, the wire (filament) gets so hot that it becomes white hot (about 2400°C) thus giving off light as well as some heat.

b Magnetic effect

As current flows, it produces a magnetic field around the wire. It is this effect which is used in electric motors, generators, solenoids and relays. It is used to create powerful electromagnets for lifting or clamping iron objects. Perhaps the most common example of the magnetic effect in use, is in an electric door bell (Diagram 9.2). The solenoids on the soft, iron, U-shaped core form an electromagnet. When the bell-push is pressed, the circuit is complete and current passes through the solenoids setting up a magnetic field. The soft iron bar on the hammer is attracted to the electromagnet. As it moves, it makes the hammer strike the gong and at the same time breaks the circuit at the contacts. The magnetic effect ceases, the spring returns the hammer to the start again and the sequence will continue until the bell-push is released.

c Chemical effect

All metals are electrical conductors and some liquids are too. Certain liquids (called electrolytes) undergo a chemical change when a current is passed through them. This is the basis of electroplating and anodising.

An electroplating bath is used to coat an object with a thin layer of metal for a special purpose, e.g. silver and chrome for appearance as well as preventing corrosion (Diagram 9.3).

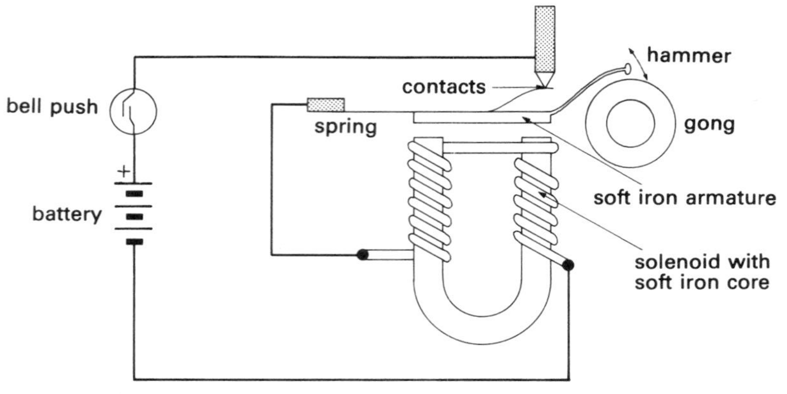

Diagram 9.2

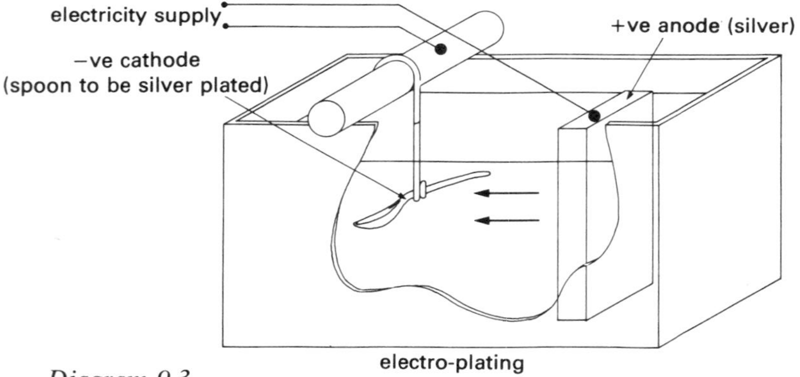

Diagram 9.3

A very similar arrangement can be used for anodising articles made in school workshops and there may be an anodising bath in your school.

9.3 Voltage (potential difference)

Electric current does not flow by itself, it has to be made to flow by an electrical pressure. The source of an electrical supply (a battery, dynamo, generator, etc.) is the electro-motive force (e.m.f.). It is this e.m.f. which creates the electrical pressure when charge flows.

Voltage (potential difference) is the electrical pressure which causes a flow rate of charge (i.e. current) in an electrical circuit. It is given the symbol V (or p.d.) and is measured in volts (also symbol V). 1 volt is the electrical pressure that causes 1 coulomb of charge to do 1 joule of work. 2 volts would cause 1 coulomb to do 2 joules of work, and so on.

$$\text{Therefore voltage} = \frac{\text{work (or energy transfer)}}{\text{charge}}$$

$$V = \frac{E}{Q}$$

But we already know that:

$$Q = It$$

$$\therefore V = \frac{E}{It}$$

$$\text{or } E = VIt$$

In words:

energy transfer = voltage × current × time

Example 3

An electric kettle is supplied with electricity from the mains at 240 V. It takes a current of 10 A for 5 min. Find the energy transfer in kilojoules.

Solution

Energy transfer = voltage × current × time
= 240 × 10 × (5 × 60)
= 720 000 J
∴ energy transfer = 720 kJ

To measure the voltage across a practical circuit we use an instrument called a voltmeter. This is connected **across** the circuit or component to find the difference in electrical pressure, i.e. voltage. An ammeter must be inserted **into** the circuit so that it can measure the current which is flowing.

Very often a comparison is made between water pipe circuits and electrical circuits (see Diagram 9.4). Comparing the two circuits we can see that the battery or source represents the pump, the ammeter

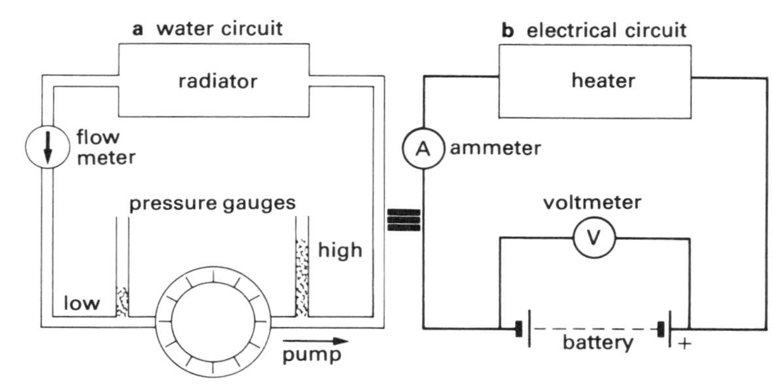

Diagram 9.4

represents the flow meter and the voltmeter represents the comparison of high/low readings on the pressure gauges. Also, the flow of current is from high pressure to low pressure. In the electrical circuit diagram the battery is shown as a series of long thin lines parallel to short thick lines. This is the usual way of showing a battery in a circuit and the long line represents the positive terminal. The current then flows from the positive terminal (high pressure) round the circuit to arrive at the negative terminal (low pressure).

9.4 Resistance

All materials conduct some electricity. Some are good and some are very poor. Those materials which are good at conducting electricity are called conductors and those which are poor are called insulators. For example, good conductors could be made of silver or copper and insulators could be made of glass, rubber or PVC.

A good conductor is one which makes little resistance to passing current. It lets current flow with very little voltage required, i.e. very little electrical pressure.

Resistance is therefore a measure of how much voltage is required to make current flow.

Consider two wires which are the same length and same diameter but made of different material. We could carry out an experiment to

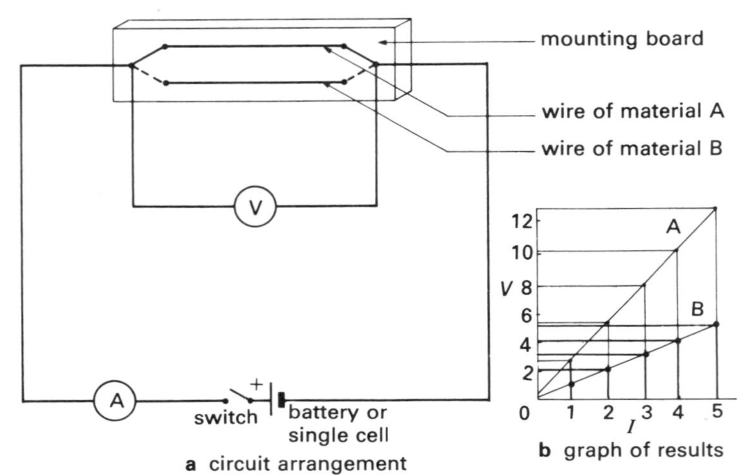

Diagram 9.5

compare the voltage required to make certain values of current flow in each conductor in turn.

The circuit would be like that shown in Diagram 9.5**a** and the graph of V against I is shown in Diagram 9.5**b**. For the two different materials, we get two different graphs, but they are both straight lines. The graphs tell us two things:

a The wire made of material A needs a bigger voltage across it than the wire made of material B to make the same current flow, i.e. the resistance of wire A is greater than the resistance of wire B.

b Because the graphs are straight line graphs, the voltage is directly proportional to the current.

$$\therefore \frac{V}{I} = \text{slope of graph} = \text{a constant}$$

This constant is the resistance (symbol R) for that wire and it will be measured in volts/amp. This unit of resistance is called the ohm (symbol Ω, the Greek letter omega).

$$\therefore \frac{V}{I} = R$$

This relationship is known as **Ohm's Law** and is usually written as $V = IR$. (Ohm's Law and the unit of resistance, the ohm, are named after the 19th-century German physicist, George Ohm.)

We could carry out further experiments to discover how resistance is affected by:

i the length of a wire;
ii the cross-sectional area of a wire (or wires);
iii the temperature of a wire.

From such experiments we would find that:

i resistance is directly proportional to length (i.e. if the length is doubled, the resistance is doubled);
ii resistance is inversely proportional to the cross-sectional area, or number of wires (i.e. if the area is doubled the resistance is halved);
iii resistance usually increases with temperature (i.e. as the wire heats up, the resistance increases).

Of all the results found by experiment, the most important result is Ohm's Law, $V = IR$.

Example 4

Four wires were tested and for each wire the voltage required for a current of 1·5 A was as follows:

Wire	A	B	C	D
Voltage (V)	6	3	9	12
Current (A)	1·5	1·5	1·5	1·5

Find the resistance of each wire.

Solution

i For wire A:

by Ohm's Law, $V = IR$ or $R = \frac{V}{I}$

$$\therefore R_A = \frac{6}{1\cdot 5} = 4\Omega$$

ii For wire B: $R_B = \frac{3}{1\cdot 5} = 2\Omega$

iii For wire C: $R_C = \frac{9}{1\cdot 5}\ 6\Omega$

iv For wire D: $R_D = \frac{12}{1{\cdot}5}\ 8\ \Omega$

Example 5

A 100 ohm resistor was connected to a source with a variable voltage control. Find the current which would flow through the resistor for voltages of: **i** 2 V; **ii** 5 V; **iii** 10 V.

Solution

i By Ohm's Law, $V = IR$ or $I = \frac{V}{R}$

for 2 volts $I = \frac{2}{100} = 0{\cdot}02$ A $= 20$ mA

ii For 5 volts, $I = \frac{5}{100} = 0{\cdot}05$ A $= 50$ mA

iii For 10 volts, $I = \frac{10}{100} = 0{\cdot}1$ A

9.5 Resistors

Electrical circuits in radios, televisions, electric train sets, etc., contain many components including resistors.

A resistor is a component which is specially made to allow a certain amount of current to pass under a certain voltage. It would not be very convenient to have numerous long lengths of resistor wire in such circuits, so the wire is insulated and wrapped round a cylindrical former. In this way it takes up less space. Such a resistor is called a wire wound resistor and there may be examples of these at school for you to look at and test.

Carbon resistors are much more common and they come in a very wide variety of sizes and resistance values. Usually they are very small, taking up very little space, and this could cause problems trying to identify different resistors, e.g. how is a 2 kΩ resistor different from a 10 kΩ resistor in appearance?

To help identify carbon resistors, they are colour-coded. Four or five bands of colour are marked on the resistors when they are manufactured so that the resistance value can be worked out using a colour-code chart for resistors (see Diagram 9.6).

It is not necessary to try to remember the colour codes, but you should be able to use the colour code chart.

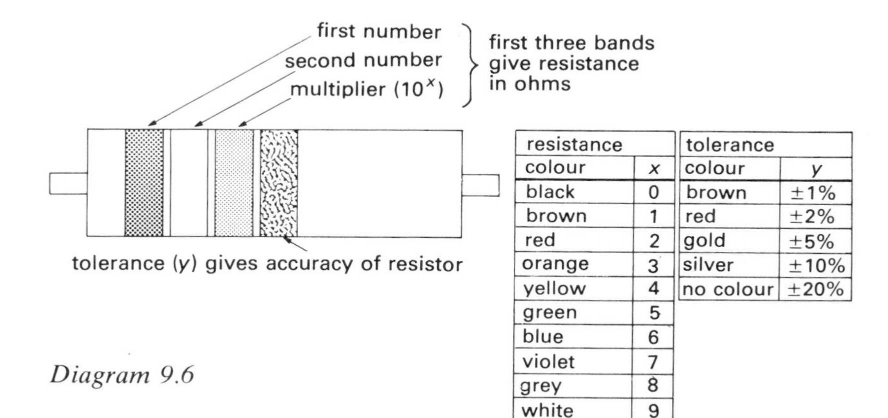

resistance	
colour	x
black	0
brown	1
red	2
orange	3
yellow	4
green	5
blue	6
violet	7
grey	8
white	9

tolerance	
colour	y
brown	±1%
red	±2%
gold	±5%
silver	±10%
no colour	±20%

Diagram 9.6

Example 6

Using the chart in Diagram 9.6, find the value of resistance for the four carbon resistors which have the following colours for their first three bands:

i red – violet – red
ii yellow – green – orange
iii red – red – green
iv orange – orange – brown

Solution

i red – violet – red = 2 – 7 – 2, from the chart
$\therefore R = 27 \times 10^2 = 2{\cdot}7\ \text{k}\Omega$

ii yellow – green – orange = 4 – 5 – 3
$\therefore R = 45 \times 10^3 = 45\ \text{k}\Omega$

iii red – red – green = 2 – 2 – 5
$\therefore R = 22 \times 10^5 = 2{\cdot}2\ \text{M}\Omega$

iv orange – orange – brown = 3 – 3 – 1
$\therefore R = 33 \times 10^1 = 330\ \Omega$

In our work we will show a resistor in a circuit simply as a rectangle, or as a short zig-zag line, with the value written beside it, as in Diagram 9.7.

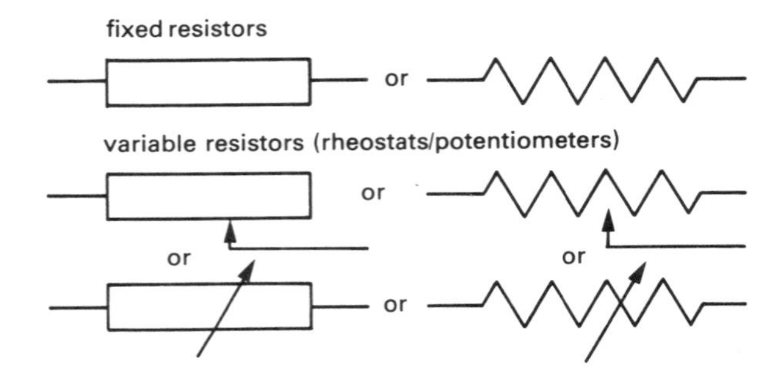

Diagram 9.7

If the value of a resistor is not stated, then it can usually be calculated from the other information given, as we shall see later.

9.6 Simple circuits

There are three essential parts in all electrical systems, a **source** of electrical energy, **conductors** for transferring the energy and the use made of the energy, the **load** (Diagram 9.8). (The word load is very often used to mean the resistance of a device, e.g. a heating element or an electric motor, etc.)

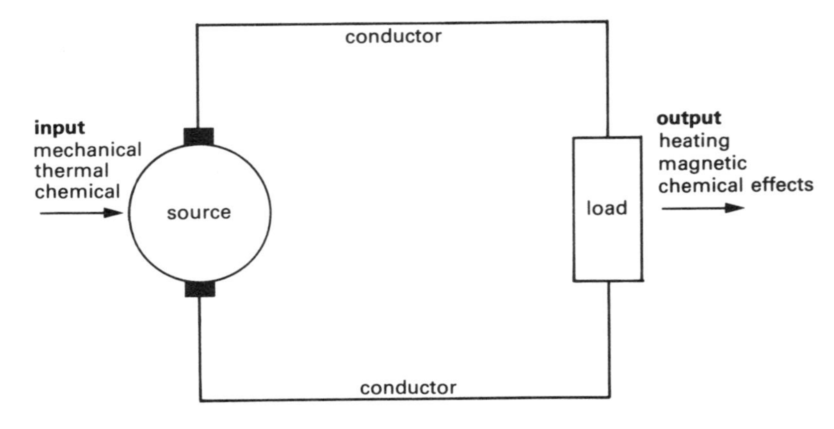

Diagram 9.8

Electrical circuits can be very simple, as in Diagram 9.4**b**, or very complicated. We will be dealing with quite straightforward circuits which will only contain a source, conductors, switches and resistance of some kind.

There are many possible arrangements for circuits. For example, Christmas-tree lights are usually arranged in a very simple way. If one bulb goes off, they all go off. Such a circuit is called a **series** circuit because each component is joined to the next one in a series or chain, and like any chain, once one link breaks, the chain is broken. A series arrangement is ideal whenever we want everything working or everything off, but if we wanted to be able to switch some things on without the others going on too, then we would have to use what is called a **parallel** circuit. Diagram 9.9 shows a series circuit and a parallel circuit for three light bulbs operated from one source. Whether to use a series arrangement or a parallel arrangement is not as simple a choice as it might first appear. Both arrangements have particular effects which we will now look at in more detail.

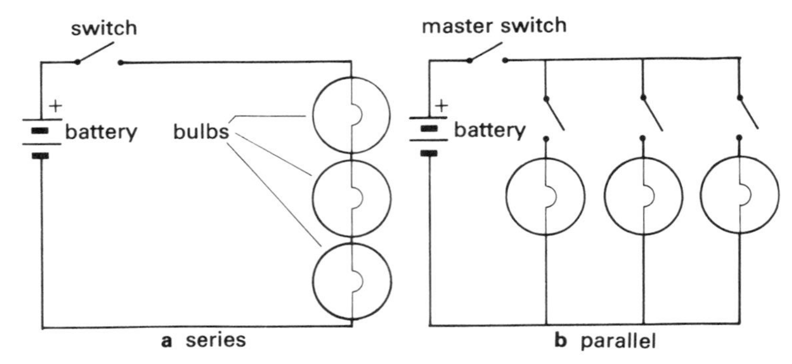

Diagram 9.9

9.7 Series circuits

If the load in a circuit is made up of two or more resistors, then there must be an overall resistance equivalent to the combined effect of the individual resistors, i.e. we could replace the individual resistors with a single resistor and produce the same effect.

Consider two resistors (R_1 and R_2) which are connected in series with an electrical source as shown in Diagram 9.10**a**. Let the equivalent resistance of R_1 and R_2 be R as shown in Diagram 9.10**b**.

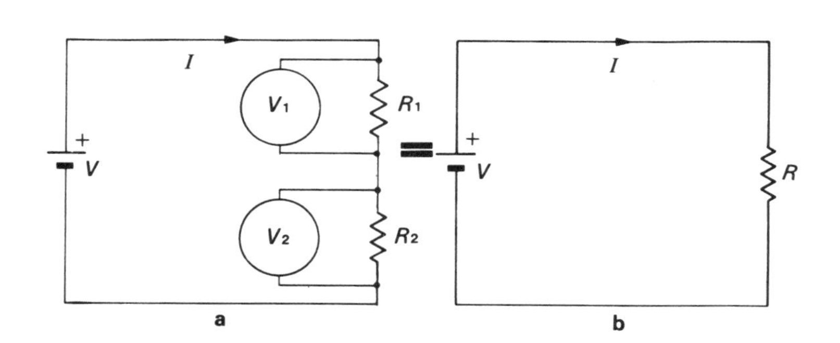

Diagram 9.10

In Diagram 9.10**a** and **b** the voltage, V, is the same and the current I is the same.

From Diagram 9.10**a**:

total voltage V = voltage across R_1 + voltage across R_2

$$\therefore V = V_1 + V_2$$

and by Ohm's Law,

voltage = current × resistance

$$V = IR_1 + IR_2$$

since the whole current, I, must pass through each resistor in turn. But from Diagram 9.10**b**:

$$V = IR$$

Therefore we have two expressions for the total voltage V, and they must be equal.

$$\therefore IR = IR_1 + IR_2$$

Since I is a common factor in all the terms, we can cancel it out.

$$\therefore R = R_1 + R_2$$

In words, the equivalent resistance of two or more resistors connected in series is the sum of the individual resistances.

Example 7

A 6Ω resistor and a 4Ω resistor are connected in series to a voltage of 4·5 V. Find **i** the equivalent resistance and **ii** the current which will flow in the circuit.

Solution

i Equivalent resistance, $R = R_1 + R_2$

$$= 6 + 4$$

$$\therefore R = 10\ \Omega$$

ii By Ohm's Law, $V = IR$

$$\therefore I = \frac{V}{R} = \frac{4\cdot5}{10}$$

∴ current, $I = 0\cdot45$ A

Example 8

Three resistors, 2Ω, 4Ω and 10Ω, are connected in series to a source. The current which is taken is 0·5 A. Find: **i** the equivalent resistance of the three resistors, **ii** the voltage across each resistor and **iii** the total supply voltage.

Solution

i Equivalent resistance, $R = R_1 + R_2 + R_3$

$$= 2 + 4 + 10$$

$$\therefore R = 16\Omega$$

ii To find the voltage across each resistor in turn we use Ohm's Law, $V = IR$.

For R_1, $V_1 = 0\cdot5 \times 2 = 1$V
For R_2, $V_2 = 0\cdot5 \times 4 = 2$V
For R_3, $V_3 = 0\cdot5 \times 10 = 5$V

iii To find the total supply voltage, we can do this in two ways:
a as the sum of the individual voltages in series

$$\therefore V = V_1 + V_2 + V_3 = 1 + 2 + 5 = 8\ \text{V}$$

or **b** as a voltage which drives the current of 0·5A through the equivalent resistance, R, of 16Ω.

$$\therefore V = IR = 0\cdot5 \times 16 = 8\ \text{V}$$

Example 9

A voltage of 4·5 V drives a current of 50 mA round a circuit which contains two resistors wired in series. R_1 has a resistance of 60 Ω, find R_2.

Solution

Equivalent resistance, $R = \frac{V}{I} = \frac{4 \cdot 5}{0 \cdot 05} = 90\,\Omega$

Also, equivalent resistance $R = R_1 + R_2$

$$\therefore R_2 = R - R_1$$
$$= 90 - 60$$
$$\therefore R_2 = 30\,\Omega$$

9.8 Parallel circuits

Consider two resistors, R_1 and R_2, connected in parallel with a source as shown in Diagram 9.11**a**. Let the equivalent resistance of these two resistors be R as shown in Diagram 9.11**b**.

From Diagram 9.11**a** we can see that the same electrical pressure, V, acts across both resistors,
i.e. voltage across R_1 = voltage across $R_2 = V$.

Also, at joint X, the current divides into two smaller currents; I_1 going through R_1 and I_2 going through R_2. After these two smaller currents have passed through their resistors, they come to joint Y where they combine again into the total current I.

$$\therefore I = I_1 + I_2$$

By Ohm's Law, $V = IR$

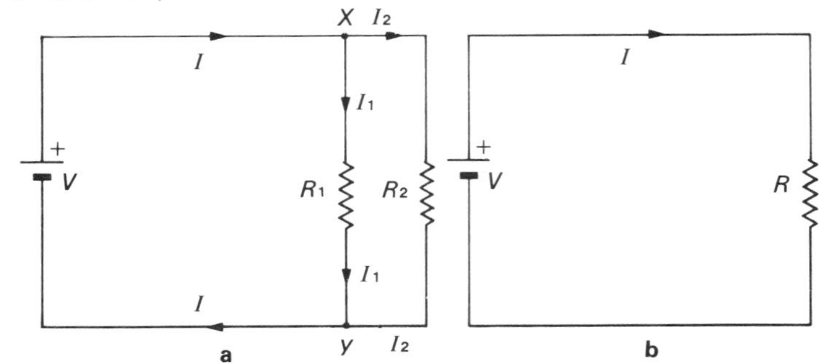

Diagram 9.11

or $$I = \frac{V}{R}$$

Therefore, from Diagram 9.11**a**:

$$I = I_1 + I_2$$
$$\therefore I = \frac{V}{R_1} + \frac{V}{R_2}$$

But from Diagram 9.11**b**:

$$I = \frac{V}{R},\ \text{where } R = \text{equivalent resistance.}$$

Therefore we have two expressions for the total current I, and they must be equal.

$$\therefore \frac{V}{R} = \frac{V}{R_1} + \frac{V}{R_2}$$

Since V is a common factor in all the terms, we can cancel it out.

$$\therefore \frac{1}{R} = \frac{1}{R_1} + \frac{1}{R_2}$$

In words, the reciprocal of the equivalent resistance of two or more resistors connected in parallel is the sum of the reciprocals of the individual resistors. (Reciprocal means 'one over . . .') Note that the value of the equivalent resistance for a set of parallel resistors will **always** be smaller than that of the smallest resistor.

Example 10

Three resistors are connected in parallel. Find the equivalent resistance when $R_1 = 3\,\Omega$, $R_2 = 4 \cdot 5\,\Omega$, $R_3 = 9\,\Omega$.

Solution

$$\frac{1}{R} = \frac{1}{R_1} + \frac{1}{R_2} + \frac{1}{R_3}$$
$$= \frac{1}{3} + \frac{1}{4 \cdot 5} + \frac{1}{9}$$
$$= \frac{3}{9} + \frac{2}{9} + \frac{1}{9}$$
$$\therefore \frac{1}{R} = \frac{6}{9}$$

To find the equivalent resistance, R, we must invert both sides of this equation.

$\therefore \frac{R}{1} = \frac{9}{6}$

$\therefore R = 1{\cdot}5\,\Omega$

Graphical method for resistors in parallel

This is a very simple method of finding the equivalent resistance for resistors in parallel, but it must be drawn accurately to give acceptable results.

Consider two resistors, $40\,\Omega$ and $60\,\Omega$, connected in parallel. Using the graphical method, we draw a base line and then, to some chosen scale, say 1 mm to $1\,\Omega$, draw two vertical lines to represent the magnitudes of the two resistances, as shown in Diagram 9.12. (The distance between them on the base line is not important – choose a convenient spacing.)

Then we join the top of each line to the bottom of the other and where these lines cross is the top of the equivalent resistance R, shown dotted in Diagram 9.12. The length of the dotted line (for R) is measured to scale and the value of R is thus found. In this case, the length of the R line is 24 mm and the scale is 1 mm to $1\,\Omega$, therefore the equivalent resistance, $R = 24\,\Omega$.

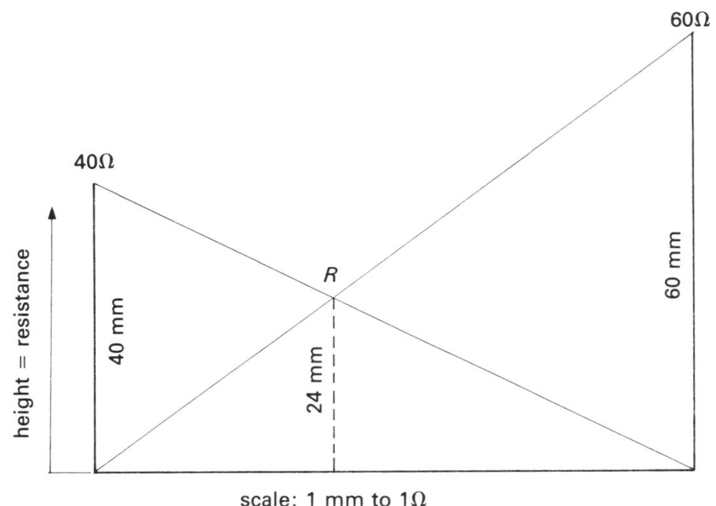

Diagram 9.12

Example 11

Find the equivalent resistance for a $24\,\Omega$ resistor connected in parallel with $30\,\Omega$ resistor.

Solution

From Diagram 9.13, the length of the R line = 26·5 mm

$\therefore$ equivalent resistance $= 26{\cdot}5 \times 0{\cdot}5$

$\therefore R = 13{\cdot}3\,\Omega$

We can check this answer, using the other method:

$$\frac{1}{R} = \frac{1}{R_1} + \frac{1}{R_2}$$

$$= \frac{1}{24} + \frac{1}{30}$$

$$= \frac{5}{120} + \frac{4}{120}$$

$$= \frac{9}{120}$$

$$\therefore R = \frac{120}{9} = 13{\cdot}3\,\Omega$$

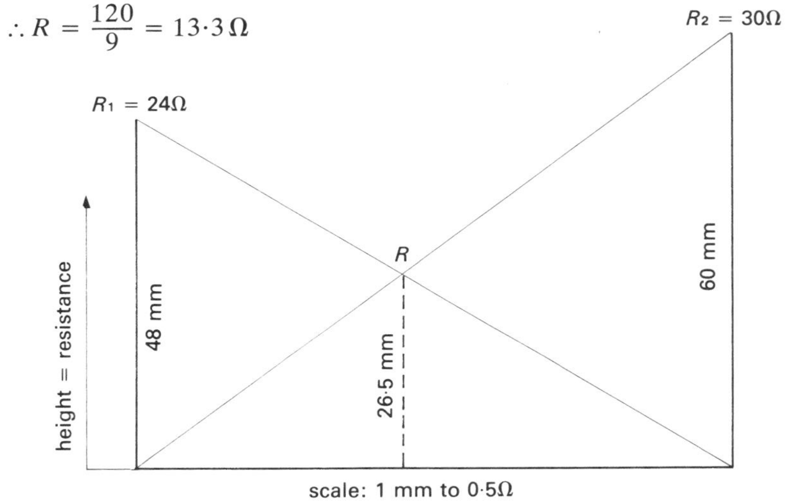

Diagram 9.13

The graphical method can be extended for three or more resistors in parallel. This is done by finding the equivalent resistance for the first two. Then, taking this as a resistor, find the equivalent resistance for this with the next resistor and so on.

Example 12

Find the equivalent resistance for the three resistors, 40 Ω, 50 Ω and 60 Ω, connected in parallel.

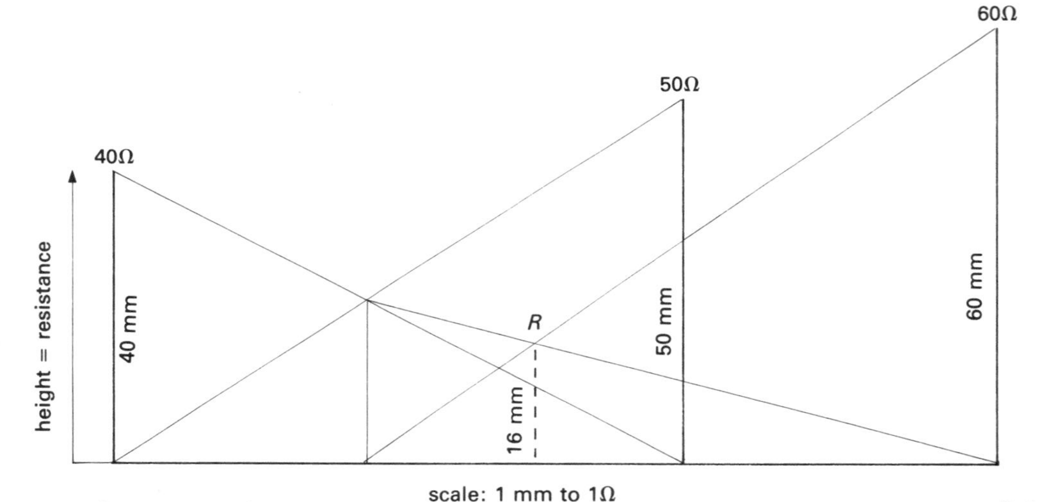

Diagram 9.14

Solution

From Diagram 9.14, the length of the R line is 16 mm.

$\therefore R = 16\,\Omega$

(Using the analytical method, we would find that $R = 16{\cdot}2\,\Omega$. Therefore the graphical method is within 2%, which, for graphical methods, is acceptable.)

Example 13

Three resistors are connected as shown in Diagram 9.15. Find the equivalent resistance of the three.

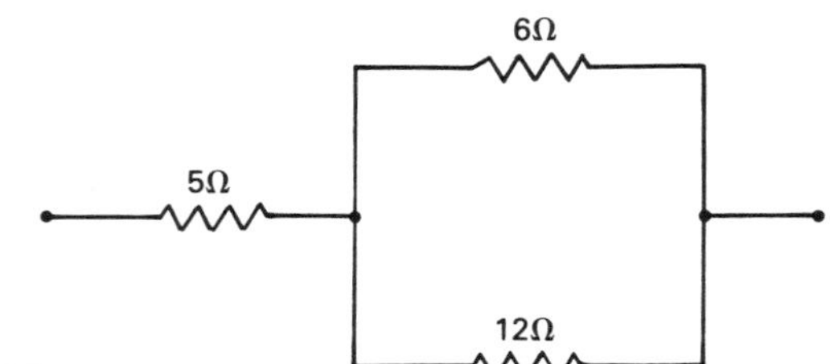

Diagram 9.15

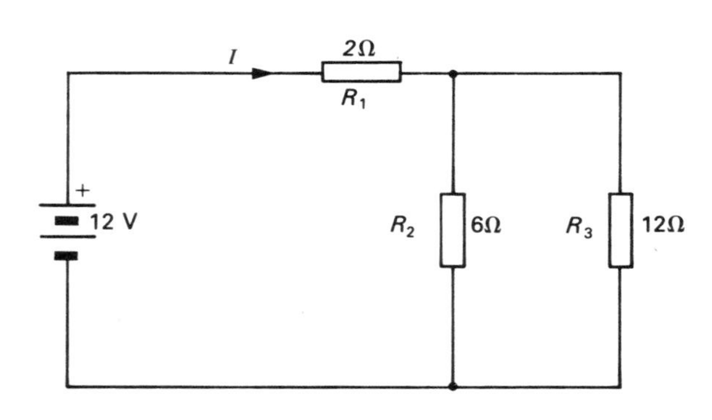

Diagram 9.16

Solution

First of all we must find an equivalent resistance for the parallel pair:

$$\frac{1}{R} = \frac{1}{6} + \frac{1}{12} = \frac{2}{12} + \frac{1}{12} = \frac{3}{12}$$

$$\therefore R = \frac{12}{3} = 4\,\Omega$$

We now have a 5 Ω resistor in series with an equivalent resistance of 4 Ω.

$\therefore$ The overall equivalent resistance $= 4 + 5 = 9\,\Omega$

Example 14

For the circuit shown in Diagram 9.16, find **i** the equivalent resistance, and **ii** the current taken from the supply.

Solution

i To find the overall equivalent resistance, we must first find the equivalent resistance for the parallel pair:

$$\frac{1}{R} = \frac{1}{6} + \frac{1}{12} = \frac{2}{12} + \frac{1}{12} = \frac{3}{12}$$

$$\therefore R = \frac{12}{3} = 4\,\Omega$$

This is in series with a $2\,\Omega$ resistor.

$\therefore$ overall equivalent resistance $= 4 + 2 = 6\,\Omega$

ii To find the current taken from the supply, we use Ohm's Law:

$$V = IR$$

$$\therefore I = \frac{V}{R}, \text{ where } R \text{ is the equivalent resistance}$$

$$= \frac{12}{6}$$

$$= 2\text{ A}$$

$\therefore$ current taken from supply $= 2$ A

Example 15

A 110 V generator supplies 5 A to the resistors shown in Diagram 9.17. Find **i** the value of R_1 in the circuit, **ii** the reading on the voltmeter shown, and **iii** the reading on the ammeter shown.

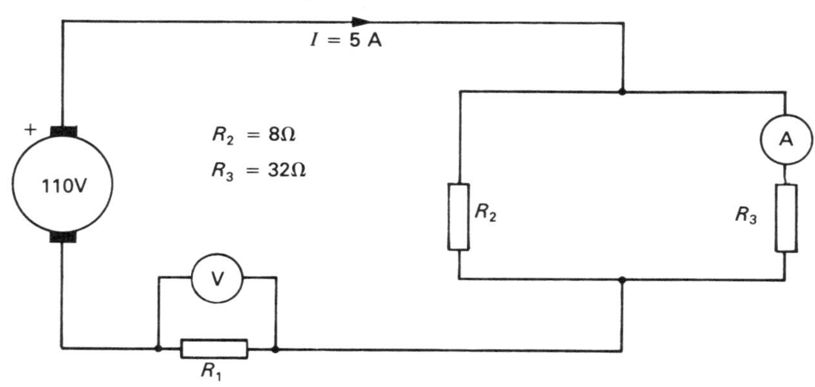

Diagram 9.17

Solution

i To find the value of R_1 we must work back from the equivalent circuit resistance. From Ohm's Law, $V = IR$

$$\therefore \text{ equivalent resistance, } R = \frac{V}{I} = \frac{110}{5} = 22\,\Omega$$

For the parallel pair of resistors:

$$\frac{1}{R} = \frac{1}{R_2} + \frac{1}{R_3}$$

$$= \frac{1}{8} + \frac{1}{32}$$

$$= \frac{4}{32} + \frac{1}{32}$$

$$= \frac{5}{32}$$

$$\therefore R = 6{\cdot}4\,\Omega$$

equivalent resistance = resistance for parallel pair + the series resistance

$$\therefore 22 = 6{\cdot}4 + R_1$$

$$\therefore R_1 = 15{\cdot}6\,\Omega$$

ii For the voltmeter: $V = IR_1$

$$= 5 \times 15{\cdot}6$$

$$\therefore V = 78\ V$$

iii For the ammeter:

the voltage across the parallel pair $= 100 - 78$

$$= 32\ V$$

By Ohm's Law $V = IR$

$$\therefore I = \frac{V}{R_3} = \frac{32}{32} = 1\ A$$

9.9 Energy and power

From the definition of voltage in section 9.3 we know that,

$$\text{voltage} = \frac{\text{energy transfer}}{\text{charge}},$$

and, $E = VIt$ (remember that energy transfer can be E or W)

Since we have the energy transfer, E, and the time taken, t, then we can find the power.

$$\text{Power, } P = \frac{\text{energy transfer}}{\text{time taken}} = \frac{E}{t}$$

but $E = VIt$

$\therefore P = VI$

We have already dealt with energy and power in Chapter 7, and we know that the unit of energy is the joule (J) and the unit of power is the watt (W).

In electrical energy transfers, there is another way of writing the units of energy.

$$\text{Power} = \frac{\text{energy transfer}}{\text{time}}$$

$\therefore$ energy transfer = power × time

Therefore, multiplying the units of power by the units of time gives units like watt-seconds and kilowatt-hours.

1 kilowatt-hour (1 kW h) is what the Electricity Boards call 'a unit' of electrical energy. Electricity bills are sent to consumers, charging them a few pence for every unit of electrical energy they have used. The electricity meters in our houses measure the number of units of electrical energy used by us.

1 unit = 1 kW h
But 1 kW = 1000 W and 1 h = 3600 s
$\therefore$ 1 kW h = $3{\cdot}6 \times 10^6$ W s or J

Example 16

How much energy is used by a 3 kW electric fire which is on for 8 hours?

Solution

Energy = power × time
= 3 × 8
= 24 kW h

Example 17

How much does it cost to boil a kettle full of water six times a day for a week? The kettle is rated at 2 kW, takes 8 minutes to boil and 1 unit of electrical energy costs 5 pence.

Solution

Power of kettle = 2 kW
total time = 8 × 6 × 7
= 336 min
energy = power × time
$= 2 \times \frac{336}{60}$
$= \frac{672}{60}$
= 11·2 kW h
cost = number of units × charge per unit
= 11·2 × 5
= 56 pence

The electrical energy which is used is actually changed into another form by the device depending on what effect is required; magnetic, heating or chemical. The energy supplied to the device is sometimes called the energy dissipated in the Load.

Example 18

Two 60 Ω bars on an electric fire are to be connected to the mains (240 V) in such a way that maximum electrical energy can be made available for converting into thermal energy. Should the bars be connected in series with each other or in parallel?

Solution

Diagram 9.18 shows the two possible arrangements.

a Series

With the two bars connected in series, the equivalent resistance is 120 Ω.

$\therefore$ Circuit current, $I = \frac{V}{R} = \frac{240}{120} = 2$ A

$\therefore$ power $= VI = 240 \times 2 = 480$ W

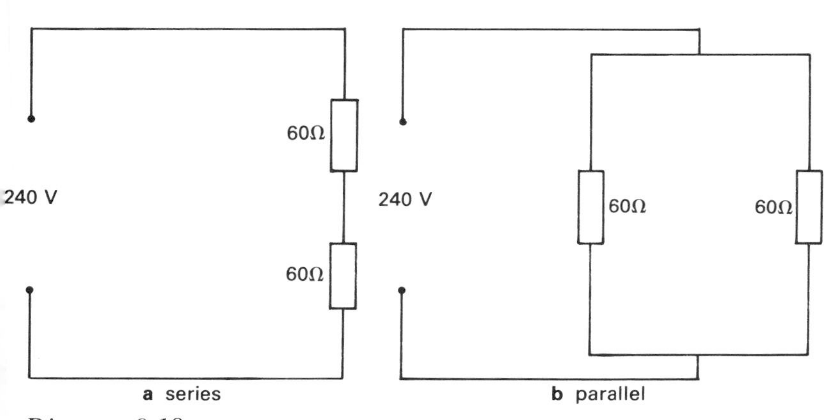

Diagram 9.18

Consider the bars to be on for 1 hour

$$\begin{aligned} \text{energy} &= \text{power} \times \text{time} \\ &= 0{\cdot}480 \times 1 \\ &= 0{\cdot}48 \text{ kW h} \end{aligned}$$

b Parallel
With the two bars connected in parallel, the equivalent resistance is found as follows:

$$\frac{1}{R} = \frac{1}{60} + \frac{1}{60} = \frac{1}{30}$$

$\therefore$ equivalent resistance $= 30\ \Omega$

$$\therefore \text{circuit current} = \frac{V}{R} = \frac{240}{30} = 8 \text{ A}$$

$$\therefore \text{power} = VI = 240 \times 8 = 1{\cdot}92 \text{ kW}$$

Consider the bars to be on for 1 hour.

$$\begin{aligned} \text{Energy} &= \text{power} \times \text{time} \\ &= 1{\cdot}92 \text{ kW h} \end{aligned}$$

Therefore to obtain the transfer of maximum electrical energy, the bars should be connected in parallel.

Example 19

For the circuit shown in Diagram 9.19, find: **i** the current, I, **ii** the potential difference (voltage) across **a** – **b** (i.e. V_{ab}), and **iii** the energy dissipated in R_5 in 10 hours.

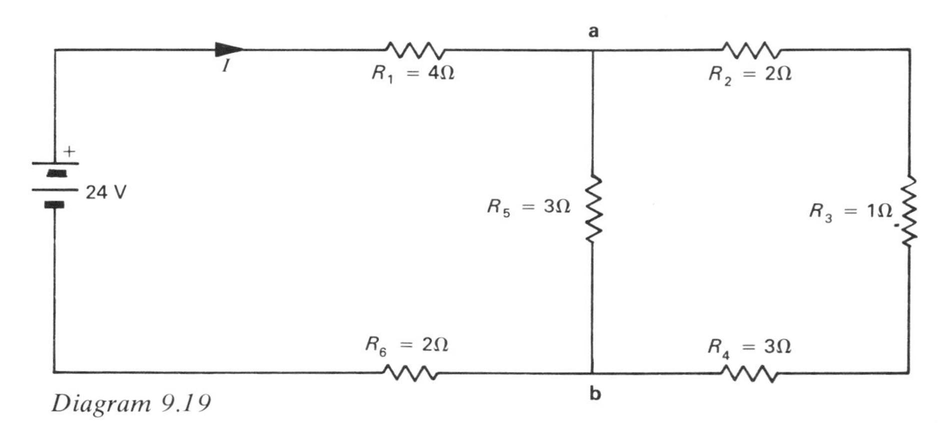

Diagram 9.19

Solution

Although the circuit may appear to be a complicated arrangement of resistors, it is in fact only a combination of series and parallel resistors.

i To find the circuit current, I, we must first find the equivalent circuit resistance for the six resistors. The resistors R_2, R_3 and R_4 are in series with each other along one branch between **a** and **b**.

$\therefore$ equivalent resistance of R_2, R_3 and $R_4 = 2 + 1 + 3 = 6\,\Omega$

This is in parallel with R_5:

$\therefore$ equivalent resistance between **a** and **b** is found by:

$$\frac{1}{R} = \frac{1}{3} + \frac{1}{6} = \frac{1}{2}$$

$\therefore R = 2\,\Omega$

Therefore we are left with three resistors in series: R_1, the equivalent resistance between **a** and **b** and the last resistor R_6.

$\therefore$ overall equivalent resistance of circuit $= 4 + 2 + 2 = 8$

$$\therefore \text{circuit current, } I = \frac{V}{R} = \frac{24}{8} = 3 \text{ A}$$

ii To find V_{ab}:
By Ohm's Law $V = I\,R$

$$\begin{aligned} \therefore V_{ab} &= I \times \text{equivalent resistance } \mathbf{a} \text{ to } \mathbf{b} \\ &= 3 \times 2 \\ \therefore V_{ab} &= 6 \text{ V} \end{aligned}$$

iii To find the energy dissipated in R_5 in 10 hours:

energy = power × time
= voltage × current × time

We know the time and the voltage across R_5, but we need to find the current through R_5.

Using Ohm's Law, $I = \frac{V_{ab}}{R_5} = \frac{6}{3} = 2$ A

$\therefore$ energy = 6 × 2 × 10 watt-hours
$\therefore$ energy = 0·12 kW h

9.10 Cells and batteries

A **cell** is the basic, single unit which develops an e.m.f. (i.e. voltage) by a chemical action. A **battery** is a combination of single cells to give the required total e.m.f. There are two types of single cell; a primary cell and a secondary cell.

In a **primary cell** (Diagram 9.20) the chemical changes which produce the e.m.f. are not reversible. When all the chemical changes have taken place and the cell no longer produces electrical energy, the cell is useless and has to be replaced by a new cell or new chemicals, i.e. a primary cell cannot be re-charged.

In a **secondary cell** (Diagram 9.21), sometimes called an **accumulator**, the chemical changes which produce the e.m.f. **are** reversible. The changes are reversed by passing a small, steady current through the cell in the opposite direction (rather like pumping water back up to a reservoir). Passing current through a secondary cell in the reverse direction is called **charging**.

A cell can only produce a limited size of e.m.f. Making a single cell physically larger means that it will give a higher current or last longer. A single Leclanché dry cell has an e.m.f. of 1·5 V. A single Secondary Cell (of the lead-acid type) has an e.m.f. of 2 V, although in practice this can vary from 1·8 V to 2·3 V. Therefore to produce a higher e.m.f., several cells must be joined in series (positive to negative) to add up to the required e.m.f., e.g. a 4·5 V battery requires three dry cells each of 1·5 V and a 12 V car battery requires six secondary cells (lead-acid) each of 2·0 V.

Internal resistance of cells

When a voltmeter is connected across the terminals of a cell (or battery) which is not supplying current to a circuit, the reading on the

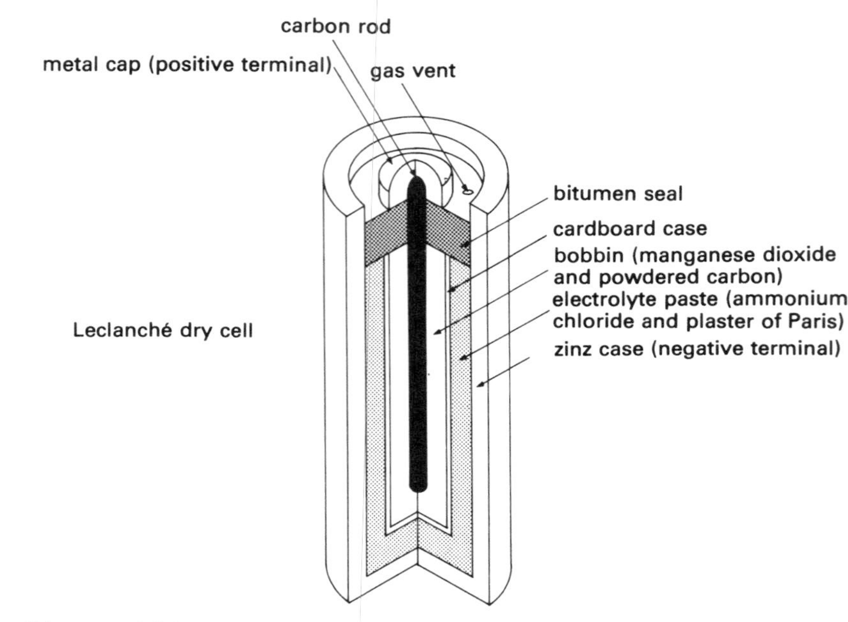

Diagram 9.20

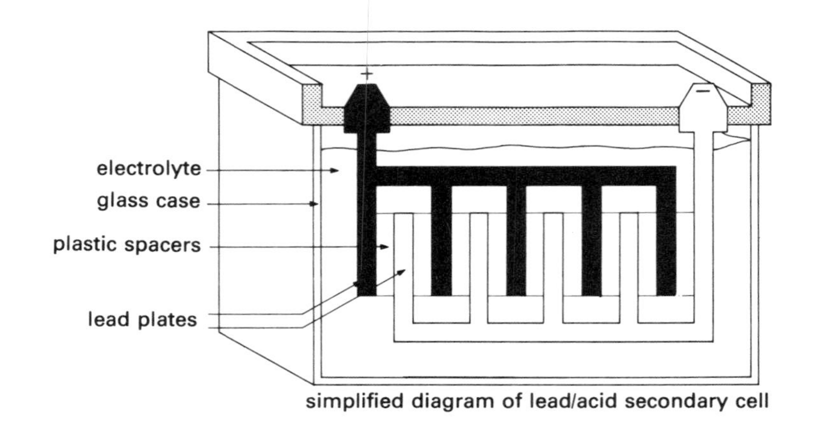

Diagram 9.21

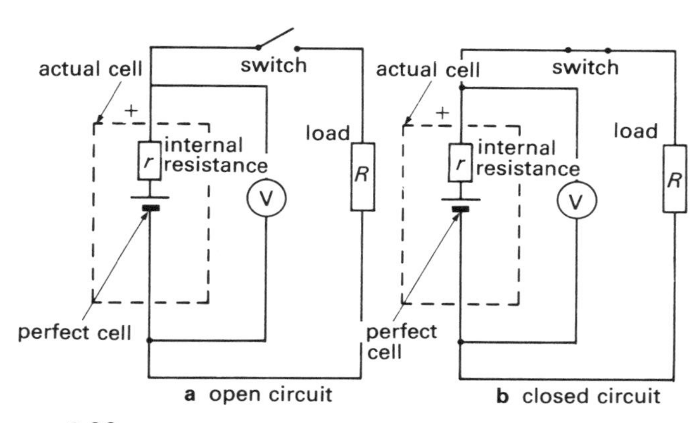

Diagram 9.22

voltmeter is the e.m.f. of that cell.

If we connect a voltmeter across the terminals when the cell **is** supplying current to a circuit we may expect to get the same reading on the voltmeter, but we would get a slightly lower reading. This is because the e.m.f. not only has to push the current around the external circuit, but it also has to push the current through the cell itself. Therefore we lose a little of the e.m.f. and what is left is the potential difference (voltage) across the external circuit. This is illustrated in Diagram 9.22.

With the switch open (i.e. off) as in Diagram 9.22**a**, there is no current flowing and therefore there is no loss of potential across the internal resistance. Therefore, for open circuit the voltage is the same as the e.m.f.

With the switch closed (i.e. on) as in Diagram 9.22**b** a current I is flowing. Therefore there will be a loss of potential because of the internal resistance. Therefore, for closed circuit:

$$V = \text{e.m.f.} - \text{potential difference across the internal resistance}$$

$$\therefore V = \text{e.m.f.} - Ir$$

From this equation we can see that small currents will cause only a small loss of e.m.f. Large currents will cause a larger loss of e.m.f. and could cause the cell to heat up.

Charging secondary cells

To charge a secondary cell a current must be passed through it in the reverse direction. We need a suitable supply voltage which is larger than the cell's voltage in order to make this happen.

It is important to charge such cells using a limited current (for lead–acid batteries this is usually less than 2A). To control the charge rate, a variable resistor can be connected in series with the cell or battery being charged, as shown in Diagram 9.23.

When a secondary cell is being charged, the electric current causes a chemical change to take place. During this change, gases are given off. These gases are hydrogen and oxygen, and under certain conditions they could form a potentially explosive mixture. Therefore, a room which is used for charging cells and batteries should have good ventilation and there should be a ban on all naked flames and on smoking in that room.

The capacity of a secondary cell is usually stated in ampere-hours. This is the amount of electrical charge that the secondary cell is capable of supplying. For example, if a car battery has a capacity of 50 ampere-hours it means that it is capable of supplying 1 amp for 50 hours, or 2 amps for 25 hours, etc.

The normal maximum rate of discharge of a secondary cell is the rate that will discharge the cell in 10 hours, i.e. 5A for a 50 A h battery; 2 A for a 20 A h battery and so on.

One way of measuring the efficiency of a secondary cell is to compare the amount of charge (i.e. ampere-hours) obtained from the cell

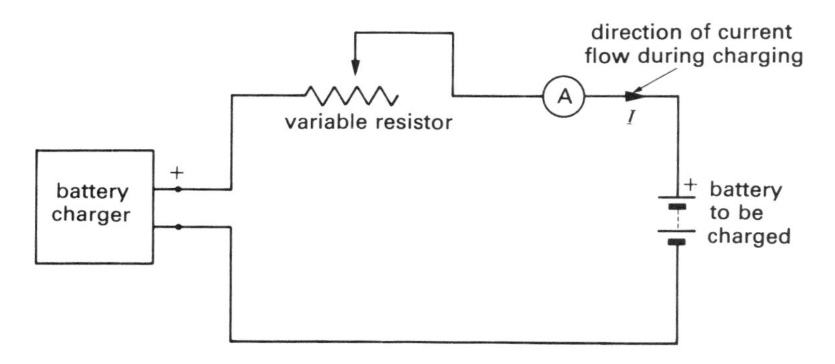

Diagram 9.23

while discharging with the amount of charge put into the cell while it was being charged.

$$\therefore \text{efficiency} = \frac{\text{discharge ampere-hours}}{\text{charge ampere-hours}}$$

This efficiency is usually about 90%. To keep a secondary cell in good condition it should not be charged or discharged too quickly and the electrolyte level must be checked. It must not be completely discharged nor left in a discharged state for any length of time. Even the best maintained battery has a limited life, but its life would be considerably shortened if it was abused or neglected.

9.11 Safety

It is important to realise that electricity can be dangerous. Injury and even death can result from electric shock. Fires can be started by electricity. In general we can say that there are two areas of safety; personal safety and appliance safety.

Personal safety

When dealing with electricity we must develop the habit of treating all sources, conductors and appliances with respect, no matter the voltage whether or not it is a.c. (alternating current) or d.c. (direct current).

The human body, like all materials, does have electrical resistance, but no two people are the same. Some may have a high resistance (about 3000 Ω) others may have a lower resistance (about 1500 Ω) depending on many factors, one of which is the moisture of the skin. These rough figures apply to direct current. For alternating current like the mains electrical supply, body resistance could be only 400 Ω. The lower the body resistance, the greater the current which will pass through the body. A fraction of one amp could be dangerous to the internal organs in the body and may even be fatal. When wearing insulating clothing (gloves and shoes) or using insulated tools, they must be clean and dry otherwise their insulating properties could be reduced.

Appliance safety

As a general rule, all appliances should be **earthed**, i.e. the frame, or any part of the appliance which normally can be touched by human hand, should be connected to earth.

In modern domestic wiring, a three wire system is used; two for distribution and one (colour coded green and yellow) for earth. The earth terminal in an appliance or in a three-pin plug is usually indicated by the letter E or the symbol for earth, ⏚.

All appliances should be **fused**. A fuse is a small length of wire which melts when the stated current is exceeded and thus breaks the circuit, effectively switching the appliance off.

There is a wide variety of fuses available and it is important that the correct rating of fuse is used for each individual appliance. New 13 amp plugs which contain a fuse are usually supplied complete with a 13 amp fuse. It would be necessary, therefore, to check that it would be correct for the appliance otherwise it must be changed; e.g. an electric iron rated at 750–890 watts and 220–240 volts would take a current of 3·4–3·7 amps. Therefore the appropriate fuse to use would be a 5 A fuse and not the 13 A fuse supplied with the plug. It should be noted that the fuse is inserted at the **line** terminal (L) in a plug to which the **brown** wire is connected. The blue wire is connected to the neutral terminal (N) in a plug (Diagram 9.24).

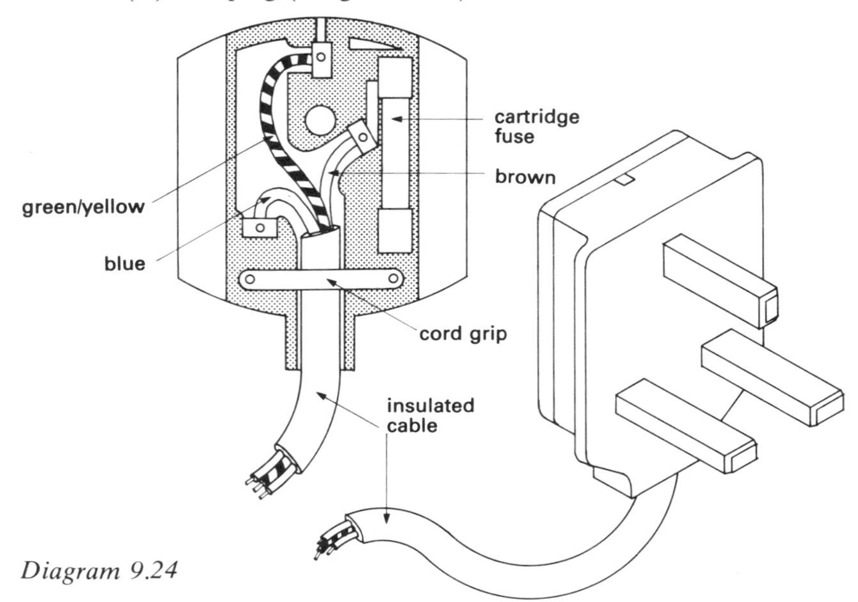

Diagram 9.24

9.12 Summary

1. Quantity of electricity is charge, measured in coulombs.
2. Charge = current × time ($Q = I\,t$).
3. The effects of an electric current are heating, magnetic and chemical effects.
4. Voltage = energy transfer per unit charge ($V = \frac{E}{Q}$).
5. Ohm's Law: voltage is directly proportional to current ($V = I\,R$).
6. Resistors can be of two types; wire wound or carbon. Carbon resistors are colour coded.
7. Resistors in series: $R = R_1 + R_2 \ldots$
8. Resistors in parallel: $\frac{1}{R} = \frac{1}{R_1} + \frac{1}{R_2} + \ldots$
9. Electrical energy transfer: $E = VIt$.
10. Electrical power: $P = VI$.
11. Electrical energy: 1 unit = 1 kW h = $3{\cdot}6 \times 10^6$ J.
12. A battery is a combination of cells.
13. A primary cell cannot be recharged.
14. A secondary cell can be recharged.
15. All cells have some internal resistance.
16. On open circuit, V = e.m.f. of a cell.
17. On closed circuit, V = e.m.f. $- Ir$.
18. Capacity of secondary cells is measured in ampere-hours.
19. Efficiency of secondary cell = $\frac{\text{discharge A h}}{\text{charge A h}}$

Exercises

1. A current of 3 amps is maintained for 40 seconds. Find the quantity of charge passed.
2. A charge of 300 coulombs is to be passed in 2·5 minutes. Find the current.
3. A current of 0·25 A passed a charge of 50 coulombs. For how long was the current maintained?
4. Diagram 9.25 shows the voltage/current graph obtained during a test on a resistor. Find: **i** the value of the resistance, **ii** the current at 1·5 V, and **iii** the voltage required to produce a current of 40 mA.

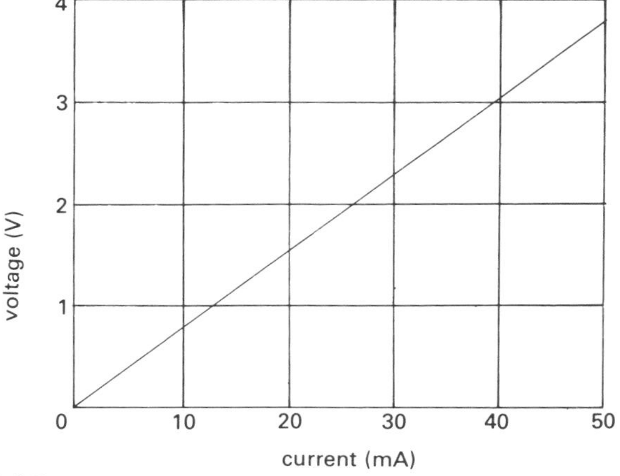

Diagram 9.25

5. Find the potential difference across a 20 Ω resistor which takes a current of 600 mA.
6. Find the current which flows through a 1 kΩ resistor when it is subjected to a potential difference of 20 V.
7. A 4·5 V battery passes a current of 5 mA through a circuit. Find the circuit resistance.
8. A current of 72 mA passes through a circuit which has a resistance of 0·5 kΩ. Find the potential difference.

9 A voltage of 9 V is applied across a 4·5 kΩ resistor. Find the current which will flow through the resistor.

10 A resistor passes a current of 30 μA when the potential difference across it is 6 mV. Find the resistance.

11 Using the colour code chart for carbon resistors in Diagram 9.6, find the values of the resistors whose first three colour bands are as follows:

i brown – brown – brown
ii brown – grey – green
iii yellow – orange – red
iv blue – grey – brown
v grey – red – orange

12 A 15 Ω resistor is connected in series with a 45 Ω resistor. What is their equivalent resistance?

13 Three 1·1 kΩ resistors are connected in series with each other. What is their equivalent resistance?

14 A circuit resistance of 1·5 kΩ is required. Two resistors are connected in series to meet this requirement. One has a resistance of 820 Ω. What is the resistance of the other?

15 A circuit has to pass 0·6 A when it has a potential difference of 12 V across it. Find: **i** the circuit resistance, and **ii** the value of resistance of a third series resistor when the other two resistors are 5·6 Ω and 6·8 Ω.

16 A 4 Ω and a 6 Ω resistor are connected in parallel. Find their equivalent resistance.

17 Three parallel resistors have values of resistance of 1·2 kΩ, 820 Ω and 2·4 kΩ. Find their equivalent resistance.

18 A 5 Ω, a 10 Ω and a 15 Ω resistor are connected together. Find the eight different equivalent resistances which could be produced depending on how the three are connected. (Note: all three must be used each time.)

19 A 2 Ω, a 4 Ω and a 6 Ω resistor are connected in parallel with each other and in series with a 3 Ω and a 5 Ω resistor which form a parallel pair. Find: **i** their overall equivalent resistance and **ii** the current taken from a battery whose potential difference is 4·75 V.

20 A 160 Ω, a 100 Ω and a 56 Ω resistor are connected in parallel across a battery whose p.d. is 8·5 V. Find: **i** the equivalent resistance, **ii** the current taken from the battery, and **iii** the current through each resistor.

21 An electric iron rated at 800 W is used continuously for two and a half hours. Find: **i** the electrical energy consumed in that period and **ii** the cost for that period if 1 unit costs 3·5p.

22 A kettle is connected to a 240 V main supply and draws a current of 10 A for 5 minutes in order to boil the kettle. If the element of the kettle is 80% efficient, find: **i** the electrical energy supplied; **ii** the energy usefully transferred to the kettle; **iii** the energy wasted; and **iv** how many times the kettle can be boiled for 60p when 1 unit costs 4p.

23 An electric fire, rated at 3 kW is connected to a 240 V source. Find the current it will take.

24 An electric cooker is protected by a 30 A fuse and is supplied by mains electricity at 250 V. If the working current should not exceed 20 A, find **i** the maximum safe power and **ii** the power which would be likely to melt the fuse.

25 The rear screen heater in a car is connected to the 12V battery and draws a current of 10 A. Find: **i** the power rating of the screen heater and **ii** the resistance of the screen heater and its connecting cables.

26 The power ratings for the lamp bulbs on a car were as follows: **i** headlamps 60 W; **ii** direction indicators 24 W; **iii** sidelamps and number plate 6 W; and **iv** interior courtesy lamp 3 W. Find the current which will be taken by each type of bulb when connected to the 12 V battery and also find the resistance of each type of bulb.

27 A room is lit for 6 hours by a 150 W, 240 V bulb. Find: **i** the current taken by the bulb; **ii** the resistance of the bulb; **iii** the energy dissipated at the bulb; and **iv** the cost, if 1 unit costs 5p.

28 For the circuit shown in Diagram 9.26, find: **i** the equivalent resistance of the circuit; **ii** the reading on ammeter A_1; **iii** the reading on voltmeter V; **iv** the reading on ammeter A_2; and **v** the reading on ammeter A_3.

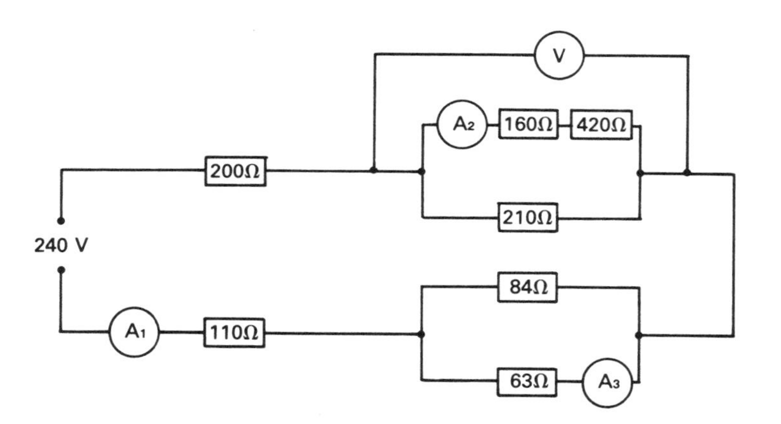

Diagram 9.26

29 The circuit shown in Diagram 9.27 is to have an equivalent resistance of 45 Ω. Find: **i** the necessary value of resistor R_A; **ii** the current taken from the supply; and **iii** the energy dissipated in R_B over a 2 hour period.

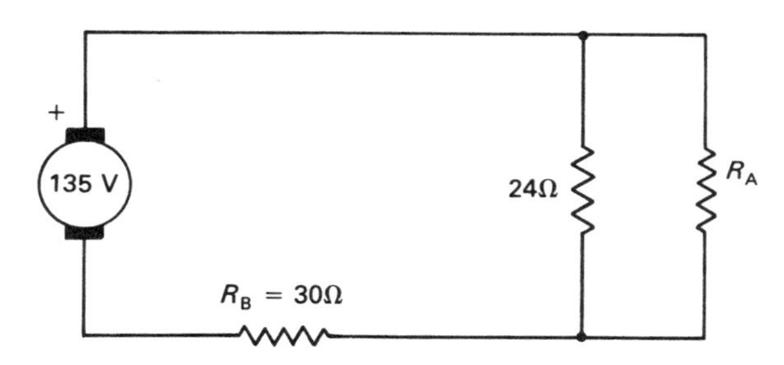

Diagram 9.27

30 Find the reading on each of the three ammeters, A_1, A_2 and A_3 in the circuit shown in Diagram 9.28.

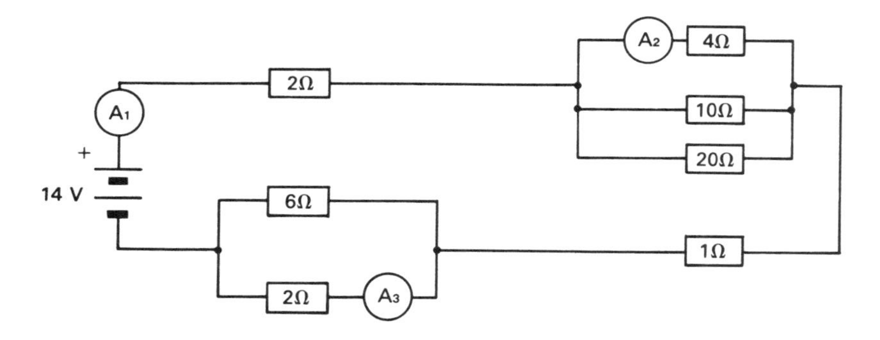

Diagram 9.28

Chapter 10
Properties of Materials

10.1 Materials

There are many different materials used today for many different purposes. To an engineer, a material is any substance which has particular characteristics. No two materials are exactly the same, but they may have similarities; they may weigh the same, they may be equally strong, they may be easily shaped but one of them may corrode very quickly. Therefore when deciding which material to use, we must consider all the things which the material will have to be able to do.

The particular characteristics which materials may have are called its **properties** and there are two kinds of properties: physical properties and chemical properties.

We will only be concerned with the more common physical properties of a limited number of materials.

10.2 Common physical properties

Before looking at particular materials it will be useful to define the properties which we will be looking for.

Density is the mass per unit volume of a substance. It is measured in kilogrammes per cubic metre (kg/m^3), e.g.

density of fresh water = 1000 kg/m^3
density of aluminium = 2700 kg/m^3
density of steel = 7800 kg/m^3

Ductility is the ability of a material to keep its strength without cracking while its shape is being altered by stretching. For example copper is ductile because it can be drawn into long, thin rods or wires. A copper bar measuring 100 mm × 100 mm can be heated, rolled and drawn into a thin round wire 20 million times longer than the original bar of copper.

Malleability is the ability of a material to keep its strength without cracking while its shape is being altered by hammering or squeezing. Gold is one of the most malleable of metals, whereas cast iron has virtually no malleability. Most metals are malleable when heated.

Toughness is the resistance of a material to breaking when it is hammered, twisted or bent. (The opposite of toughness is brittleness.)

Hardness is the resistance of a material to being cut, permanently dented or scratched. Hardness is not really an independent property, it is related to poor ductility and tensile strength.

Elasticity is the tendency of a material to return to its original size and shape after its size or shape had been altered by a force, e.g. a rubber ball can be squeezed almost flat and yet, when it is released, it springs back to its original size and shape. All materials have some elasticity, even concrete.

Electrical conductivity is a measure of how good the material is at conducting electricity. A material with a high electrical conductivity means low resistance, i.e. a good conductor like copper or aluminium. Low electrical conductivity means very high resistance, i.e. a good insulator like most plastics or rubber.

Thermal conductivity is a measure of the rate at which heat is conducted through the material, e.g. kitchen pans are made from aluminium or copper because they conduct heat very quickly. Their handles may be of wood or hard plastic because they do not conduct heat very well – they could be called thermal insulators.

10.3 Ferrous metals

These are metals which contain iron. Pure iron has few uses, but when small amounts of other substances are mixed with iron, a wide range of useful materials can be produced with varying properties. Carbon is the most common additive and this turns the iron into **steel**. The precise amount of carbon determines some of the properties of steel. In general, the carbon content is given as a percentage. Certain ranges are given particular names and have particular uses. Some of these are tabulated below.

Type of steel	Carbon content	Uses of steel
Low carbon steel	0 to 0·1%	Tubes, wire, tin-plate. Good for welding.
Mild steel	0·1 to 0·3%	Structural steel beams, ship building, general engineering (car bodies, components, gates, etc.). Good for welding.
Medium carbon steel	0·3 to 0·6%	Forgings, axles, shafts, springs, hammers and certain other tools.
High carbon steel	0·6 to 0·9%	Special springs, drills, chisels and general cutting tools.
Tool steel	0·9 to 1·2%	Axes, picks, files, razors, ball-bearings.

At the low carbon end of the scale, the steel is relatively soft and ductile. As the carbon content increases, the hardness of the steel increase, but loses some ductility and toughness. At the maximum carbon content (1·2%), the steel has extreme hardness but tends to be brittle. The hardness is useful for metal cutting tools, hence its name – tool steel.

Steel, like most metals, will conduct heat and electricity, but it is not a good conductor. The range of favourable properties for engineering purposes has made steel very important in our present society. However, it's use is so common, that steel tends to be taken for granted.

Its ease of being worked along with its toughness, makes it suitable for domestic uses too, like tin-cans, cooker frames, furniture, etc. Electric door bells, relays and electro-magnets depend on the magnetic properties of iron (a very important property which should not be overlooked).

Cast iron also contains carbon, but in higher percentages (3 or 4%). Such a high carbon content makes cast iron very hard. It has low strength in tension and is extremely brittle.

The hardness of cast iron makes it most suitable for surfaces which must not wear away quickly like brake drums, clutch plates and the slides on machines. Low grade cast iron can be used for rainwater pipes, gutters and other general or ornamental castings. Cast iron, like steel, is a relatively poor conductor of heat and electricity.

10.4 Non-ferrous metals

These are metals or mixtures of metals, which contain no iron. There are many metals in this category: aluminium, copper, lead, tin, zinc, silver, gold, etc., and all are non-magnetic. We will only look at the first two here.

Aluminium

In its pure state, aluminium is light-weight, soft, ductile a very good conductor and corrosion-resistant, but it has poor hardness properties, limited strength and is non-magnetic.

Its favourable properties make it very useful for electric power cables (where it is strengthened with a steel core) and structures in chemical plant. It makes a good light and heat reflector and its lightness makes it ideal in the aircraft industry.

Aluminium materials combine with other substances to make more materials, with slightly different properties.

Copper

For more than 5000 years, copper has been used by man. In its purest form, copper is a better conductor of heat and electricity than any other substance except silver. As a result, the electrical industry uses about 60% of all the copper produced, mainly to make electric wire. Other commonly used copper products are pots and pans, water pipes in the home, jewellery and ornamental articles. Such uses reflect other properties of copper. It is corrosion-resistant, extremely ductile, malleable and also of quite high density (it is heavier than steel).

Like aluminium, copper combines with other substances to make materials which are also very common and useful in other ways.

10.5 Alloys

An alloy is a mixture of metals or of a metal and some other substance. Most alloys contain a large amount of one particular metal and very small amounts of several other metals. The metal of which there is most is called the **base** of the alloy.

Often the disadvantage of a pure metal can be overcome by creating the right alloy – adding the right amount of the right substance. There are a great many alloys, all with particular properties and put to particular uses. (Steel, of course, is an alloy of iron and carbon.)

Duralumin is an aluminium and copper alloy. Pure aluminium is soft and ductile, but when alloyed with 4% copper it forms duralumin which is harder, slightly heavier, and almost as strong as steel. Duralumin is the best of a number of aluminium alloys and is used in structural members of all kinds, engine cylinder heads, propellors, pistons, etc.

Brass is a copper and zinc alloy. When zinc is added to copper it increases its hardness and reduces its toughness. The proportion of copper to zinc has a wide range from 85% to 15% (copper to zinc) to 55% to 45%. Each kind of brass has its own name and its own uses, but in general we can say that the greater the amount of zinc in the alloy the harder and the more brittle it becomes.

Common brass (a medium combination of copper and zinc) retains the malleability of copper, while being harder and stronger than pure copper itself.

The uses of brass are many and varied. It can be used for cheap jewellery, plumbing accessories, water tubes and simple bearings although bronze (a copper-tin alloy) would be a more suitable bearing material.

10.6 Plastics

All plastic materials are synthetic materials, i.e. they are man-made. The names of the different plastics describe the chemicals from which they were made. Some of the names are very long tongue-twisters and many plastics have had their full names shortened, e.g. polyvinyl chloride is shortened to PVC which is a well known plastics material. A real tongue-twister is polymonochlorotrifluoroethylene, abbreviated to PCTFE. Plastics as a group, covers an ever increasing range of materials, but they can be divided into two separate families of plastics depending on how they react when heated: thermoplastics and thermosetting materials.

Thermoplastics behave rather like wax. When heated they become very soft and can melt. They can be poured or moulded into almost any shape and when cool, they harden and have the same strength and other properties that they started with. This procedure can, in theory, be repeated many times – re-heating and re-forming thermoplastics materials.

The properties of thermoplastics are numerous, different plastics having special applications. They can be quite rigid or soft and flexible, coloured or transparent, but all thermoplastics are good electrical insulators.

The following examples are only a few of the thermoplastics to illustrate the range.

1 Acrylic plastics are resistant to weathering and chemical attack. They are very strong and easily coloured. They are used for optical lenses, lamp lenses for cars, outdoor signs, soles of shoes, aircraft canopies etc.

2 Nylon is a tough, strong, springy plastic which wears very well and has good electrical insulating qualities. It is used for bearings, gear wheels, toothbrushes and, in fibre form, for fabrics and stockings.

Plastic pipe being extruded, with an operative checking wall thickness

Objects made from assorted plastics

3 Polystyrene is very lightweight and cheap. It is used for toys, electrical insulation and, in its expanded form, for decorative ceiling tiles and packaging for breakables.

4 Polyvinyl chloride (PVC) is strong and flexible, easily coloured and resists scratching. It is used for water pipes, packaging, gramaphone records, waterproof clothing and insulation of electrical cables.

Thermosetting materials can be melted only once. During manufacture, heating the powder or granules causes an irreversible chemical change rather like boiling an egg. Once it has been heated and cooled it is permanent and cannot be changed back by re-heating. Thermosetting materials are good insulators both thermal and electrical and tend to be hard and rigid.

Once again, the following examples are only a few of the thermosetting materials to illustrate the range.

1 Alkyd resin has good resistance to heat and to a wide range of chemicals. Its electrical insulating properties make it suitable to use for electrical components and bases for electronic tubes. But its greatest use is in paints and enamels to provide heat and chemical resistant finishes.

2 Epoxy resin is water and weather resistant, quick hardening and strong. It is used in reinforced plastics but its best known use is in adhesives.

3 Polyester resin is also strong and quick hardening and is easily moulded at low pressure. It is used for boats, cases and tool boxes, car bodies and domestic furniture.

The development of polyester and epoxy resins has largely been for glass reinforced plastics (G.R.P.). These materials are very strong for their weight and are stiffer than other plastics.

The range of uses of both families of plastics is increasing year by year. Plastics are being used where only steel used to be suitable. Other metals are being replaced by plastics too and it is quite possible that, before long, plastics as a group will be the most used material.

10.7 Work hardening and heat treatment

When the shape of a metal is being changed by squeezing, hammering, bending, twisting, cutting, etc., the metal becomes more and

Canoe slalom racing

more resistant to the change, i.e. it loses toughness and ductility and gains hardness because it has been worked. Hence the name **work hardening**.

Sometimes this is very desirable. Steel bars, when they are produced, are very often rolled between high pressure rollers to squeeze the bars to the final size and to give the steel some hardness. Also with articles made from ductile materials like copper, the finished article is often hammered lightly all over to make it work hardened and resistant to further deformation.

However, work hardening is often an undesirable happening. For example, beating copper into a shape produces work hardening in the copper long before the shape is finished. Further hammering would make the copper split or crack unless the hardness could be removed. This is done by using a heat treatment.

Annealing is a very necessary and important heat treatment whereby metals have their hardness removed and their original soft condition restored. It is a very simple process. The metal is heated (usually to a dull red heat) and allowed to cool slowly. When the metal is cold, work can proceed again until further work hardening makes it necessary to anneal the work-piece again.

With steels there are a number of different heat treatments. Instead of simple annealing, as already described, there is steel annealing and normalising of steel; both rather similar, but used for different reasons and in any case outwith the scope of this course. It is perhaps worth mentioning one particular method of treating medium to high carbon steels. **Hardening and tempering** is a well known, well used process (or series of processes) and is common in school engineering workshops. When making articles out of carbon steels, the steel must be soft enough to be shaped, cut, filed, etc., but the article may be a cutting tool, for example a chisel. This must have a very hard cutting edge. To change the soft steel into a harder steel, the steel is heated (as for annealing) but then quenched by plunging into cool water or oil. Thus the **hardening** stage is completed. However, this will be too hard and very brittle, so some of the hardness is removed by applying a little heat and then quenching again. Removing some of the hardness in this way is called **tempering** and it restores some toughness to the steel.

Varying degrees of hardness are required for different tools or machine parts and this is done at the tempering stage. However, mild steel is so low in carbon content that heat hardening has no effect on it.

10.8 Selection of materials

In material selection we will be concerned only with 'fitness for purpose'. Therefore there could be many correct solutions. When faced with a choice we must ask ourselves questions about the importance of certain properties of the possible materials in relation to the component or article. Questions like:

Is weight important?
Must it be ductile or malleable?
Has it to stand up to hammering or bumping?
Is it to insulate (thermally or electrically)?
Is corrosion or contamination important?

By carefully answering questions like these, we can then select those materials which will satisfy the requirements, thereafter choosing the one which suits most requirements best.

Discussion points

It would be almost impossible to give detailed examples, but some components are listed as discussion points or starters. Answers have been deliberately omitted.

For each item, select appropriate materials:

1. A dustbin
2. A fireguard
3. A lightweight dining chair
4. A frying pan
5. A canoe
6. A machine vice
7. An electrical screw driver
8. Window frames
9. An extension cable
10. A power saw table
11. The reflector on a radiant fire
12. A watch face
13. An infant high chair
14. The component parts of an electric fan heater
15. An electric soldering-iron

Chapter 11

Strength of Materials

In Chapter 10 we dealt with properties of materials, some particular materials and their uses. We will also have carried out practical tests to compare various materials in a general way.

The next stage is to work out the size that a component must be so that it will be strong enough to do the job it is intended for. Before we can work out sizes we must know how the material will behave when it is subjected to a load.

In days gone by, design engineers were not very sure of the strength of the materials they were given to use and many of the structures they designed (particularly steel bridges) were very much stronger and heavier than they really needed to be. More recent structures show

The Forth rail bridge with the road bridge behind

The wreck of the Melbourne Westgate bridge after it had collapsed

how engineers have become very good at producing good quality materials and at calculating the loads which structures will have to bear. The results are usually simple, elegant, stream-lined and efficient. However, we still do not know all the answers and unpredictable things can happen, particularly when a brand new method of constructing things is being tried.

11.1 Behaviour of a material under load

Before a material is actually used in a structure, engineers test a sample or specimen of the material to see how it reacts to being stretched or compressed or twisted, etc.

The most common type of test on a specimen is a tensile test, where various increasing loads are applied to the specimen to stretch it until it breaks. A careful note is kept of what load produced what extension and also what load made the specimen break.

All test specimens are very carefully made to exact sizes and lengths to fit into special testing machines.

With the information of the dimensions of the material, the loads and the extensions, the design engineer can work out the sizes for his new structure so that it will carry the load that it is meant to.

In schools there is usually a machine for doing tests like these, called a **tensometer**, which gives readings of the load and the corresponding extension. There are also samples or specimens of various materials to test to destruction, as it is called. The shape of a typical specimen is shown in Diagram 11.1.

When a test is being carried out, the specimen seems to stretch equal amounts for equal increases in load. Then as the load increases, the amount of stretching becomes unpredictable and the specimen may become slightly narrower at one part along its length. Shortly after the specimen has developed this **waist**, it will break, usually with an impressive bang.

Diagram 11.2 shows the kind of stages the specimen will go through. Different materials behave in different ways and as many different materials as possible should be tested in this way.

By carefully recording corresponding loads and extensions, a graph can be drawn of Load (F) against extension (x).

Diagram 11.3 shows the kind of graphs which would be obtained after doing tensile tests on specimens of cast iron, mild steel and copper. All the graphs start off with a straight line portion which may be long or short.

The copper specimen seems to stretch for relatively little load and to go on stretching, which is what we would expect a ductile material to do. However, it is interesting to do the same test on another copper specimen which has been work hardened first.

The cast iron specimen has relatively little extension for a very large load before fracture occurs. Again we would expect something like this for a brittle material.

The mild steel specimen has a most interesting graph; a steep straight slope, a short level stage, a curving climb to the highest point and a curving drop to the fracture point after a reasonable extension. This pattern is typical for all steels and it is worth looking at in a little more detail.

A tensometer

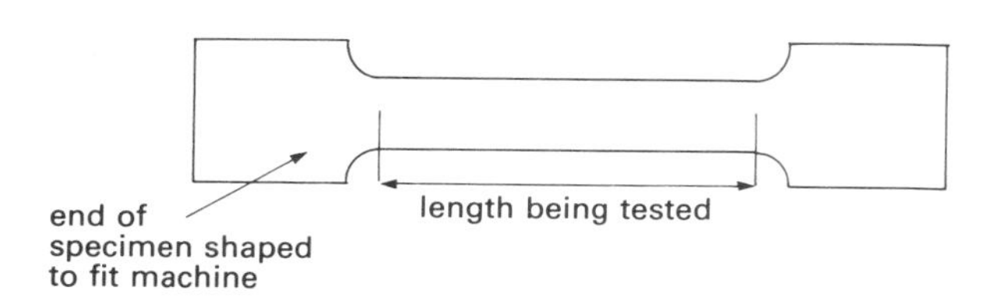

Diagram 11.1

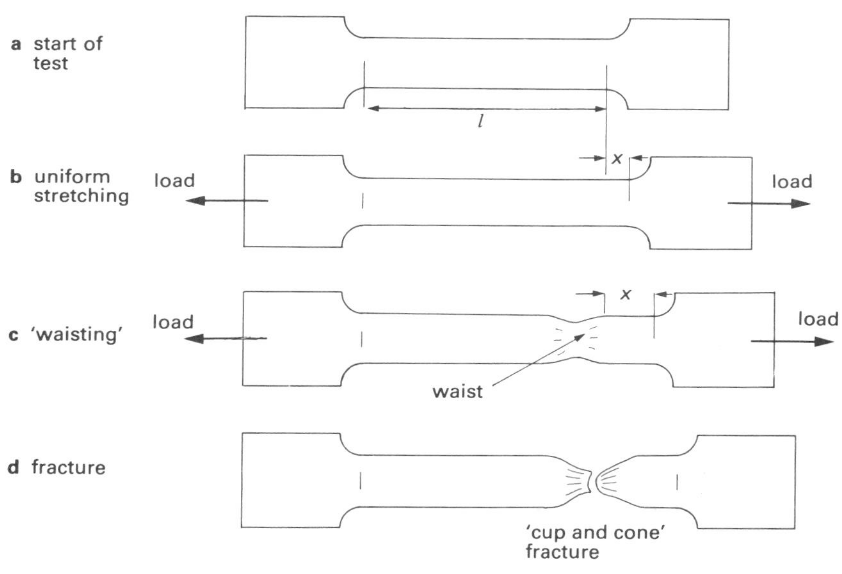

Diagram 11.2

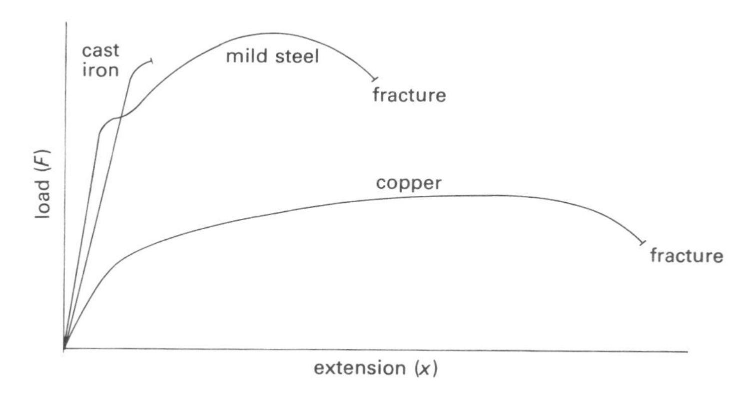

Diagram 11.3

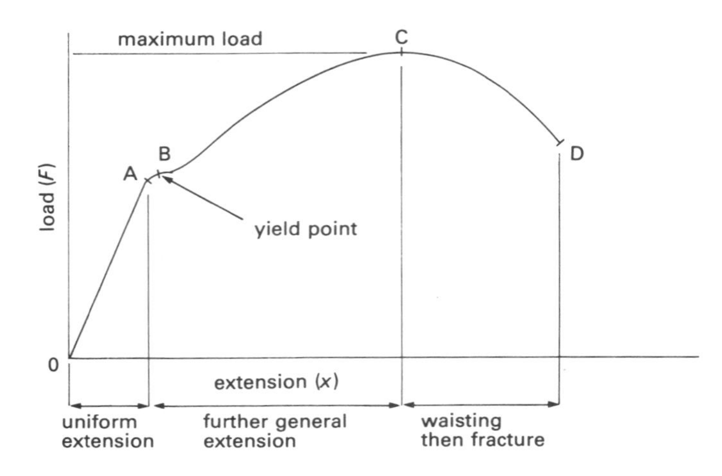

Diagram 11.4

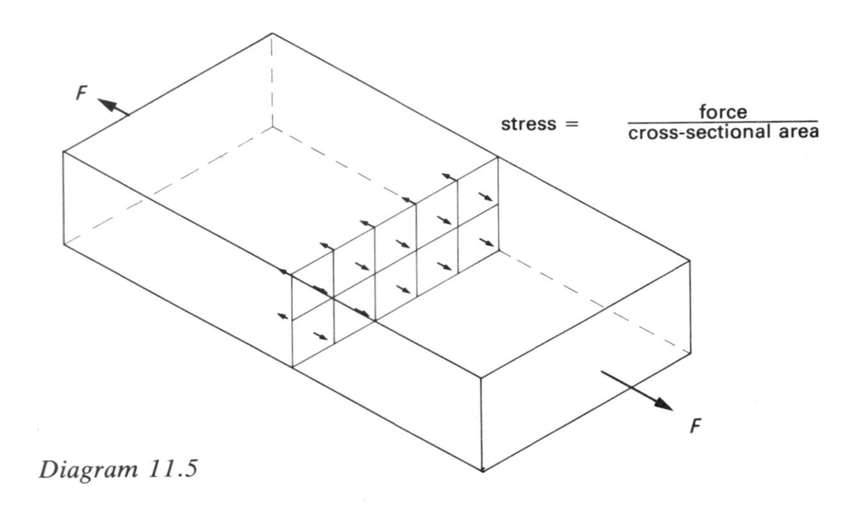

Diagram 11.5

Typical load/extension graph for steel

Up to point A on the graph, the steel extends uniformly with the increasing loads. At point B, the steel suddenly yields, i.e. it stretches on its own without any increase in the load. Then it seems to harden and further general extension can take place with increasing loads up to the maximum load at point C. At this point the specimen develops a waist (i.e. it becomes narrower at some point) and keeps stretching, even though the load gets less, until it inevitably fractures at point D. The straight line portion of the graph (from O to A) is called the elastic range and point A is usually called the elastic limit. If the test is stopped at any stage up to point A, the elastic limit, then no permanent damage will have been done and should the load be removed, the specimen will go back immediately to its original size and shape. Loading beyond point A will do permanent damage to the specimen, i.e. it will not go back to its original size and shape.

As we have seen, every material will have a straight line portion at the beginning of its load/extension graph. Therefore every material has some elasticity no matter how small.

Hooke's Law states that when a material is loaded in tension or in compression, the changes in length will be directly proportional to the loads, provided the loading is within the elastic range for that material.

i.e. change in length $\propto$ load

11.2 Stress

When a force is applied to one end of a bar, it is transmitted to the other end by the material that the bar is made of.

At any cross-section of the bar, the total force is being transmitted by that cross-section, as illustrated in Diagram 11.5.

The amount of force which is transmitted by unit area of the cross-section is called the stress and is given the symbol σ (sigma). Let the cross-sectional area be A.

$$\text{Stress} = \frac{\text{load}}{\text{cross-sectional area}}$$

or, in symbols, $\sigma = \dfrac{F}{A}$

The units of stress are units of force per unit area, e.g. N/m^2, kN/m^2, MN/m^2 or N/mm^2. In S.I. units, 1 newton per square metre is called 1 pascal ($1 N/m^2 = 1$ Pa).

$$\text{Stress} = \frac{\text{load (in newtons)}}{\text{cross-sectional area (m}^2)}\ \text{Pa}$$

Note

1 Pa is a very small unit indeed. Imagine a steel column whose cross-section measures 1 m by 1 m. If we placed an average sized eating apple on the top of such a column, it would produce a stress in the column of 1 Pa. You will find, therefore, that when working out stress you will have to use factors of 10, like 10^3, 10^6 or even 10^9 for kPa, MPa and GPa.

Example 1

A bar is subjected to a tensile load of 200 N. The cross-section of the bar measures 20 mm by 5 mm. Find the stress in the bar.

Solution

$$\begin{aligned}
\text{Cross-sectional area of bar} &= 20 \times 5 \\
&= 100\ \text{mm}^2 \\
&= 100 \times 10^{-6}\ \text{m}^2 \\
\text{Load on bar} &= 200\ \text{N} \\
\text{Stress in bar} &= \frac{\text{load (N)}}{\text{cross-sectional area (m}^2)} \\
&= \frac{200}{100 \times 10^{-6}} \\
&= 2 \times 10^6\ \text{Pa} \\
\therefore \text{stress in the bar} &= 2 \times 10^6\ \text{Pa or 2 MPa}
\end{aligned}$$

Example 2

A steel wire has a diameter of 0·8 mm and on the end of the wire hangs a mass of 14 kg. Find the stress in the wire.

Solution

$$\begin{aligned}
\text{Cross-sectional area} &= \frac{\pi d^2}{4} \\
&= \frac{\pi \times 0{\cdot}8 \times 0{\cdot}8}{4} \\
&= 0{\cdot}503\ \text{mm}^2 \\
&= 0{\cdot}503 \times 10^{-6}\ \text{m}^2 \\
\text{Load} &= mg \\
&= 14 \times 9{\cdot}81 \\
&= 137\ \text{N}
\end{aligned}$$

$$\begin{aligned}
\therefore \text{stress} &= \frac{\text{Load (N)}}{\text{cross-sectional area (m}^2)} \\
&= \frac{137}{0{\cdot}503 \times 10^{-6}} \\
&= 272 \times 10^6\ \text{Pa} \\
\therefore \text{stress in the wire} &= 272 \times 10^6\ \text{Pa or 272 MPa}
\end{aligned}$$

Example 3

The stress in a bar is not to be greater than 140 MPa and the cross-section of the bar measures 10 mm by 40 mm. What is the largest allowable load?

Solution

We know that:

$$\text{stress} = \frac{\text{load}}{\text{cross-sectional area}}$$

Therefore if we know what the maximum stress is and we know, or can calculate, the cross-sectional area, then we can find the maximum load.

$$\begin{aligned}
\text{Load} &= \text{stress} \times \text{cross-sectional area} \\
&= (140 \times 10^6) \times (10 \times 40 \times 10^{-6}) \\
&= 140 \times 400 \\
&= 56000\ \text{N or 56 kN}
\end{aligned}$$

$\therefore$ largest allowable load is 56 kN

Example 4

A tow-bar for a lorry has to be capable of exerting a pull of 10 kN. The material used for the tow-bar must not have a stress greater than 80 MPa for safety reasons. Find **i** the necessary cross-sectional area of the tow-bar in mm² and **ii** the diameter of the tow-bar if it is to have a circular cross-section.

Solution

$$\begin{aligned}
\textbf{i}\ \text{Stress} &= \frac{\text{load}}{\text{cross-sectional area}} \\
\therefore \text{cross-sectional area} &= \frac{\text{load (in N)}}{\text{stress (in Pa)}} \\
&= \frac{10 \times 10^3}{80 \times 10^6} \\
&= 0{\cdot}125 \times 10^{-3}\ \text{m}^2 \\
&= 0{\cdot}125 \times 10^{-3} \times 10^6\ \text{mm}^2 \\
\therefore \text{cross-sectional area} &= 125\ \text{mm}^2
\end{aligned}$$

ii For a circle, area $= \frac{\pi d^2}{4}$

$$\therefore \frac{\pi d^2}{4} = 125$$

$$\therefore d^2 = \frac{4 \times 125}{\pi}$$

$$= 159$$

$$\therefore d = \sqrt{159}$$

$$= 12{\cdot}6 \text{ mm}$$

$\therefore$ diameter of tow-bar = 12·6 mm (minimum)

11.3 Strain

When we apply a load to a bar, it either extends or is compressed. This we know from our earlier work on load/extension graphs. The actual size of the extension does not really tell us very much about what is happening to the material. For example, an extension of 10 mm on a steel bar which was originally 1 m long is quite a change. But an extension of 10 mm on a steel bar which was originally 10 m long is not as much of a change.

To get an idea of how much the whole bar has stretched (or compressed) we take the ratio of the change in length to the original length, but both must be measured in the same units, e.g. metres. This ratio is called the **strain**. It is a decimal fraction and has no units. The symbol for strain is ϵ (epsilon).

For a change in length we use the symbol x and for the original length we use the symbol l.

$$\text{Therefore strain} = \frac{\text{change in length (in m)}}{\text{original length (in m)}}$$

or, in symbols, $\epsilon = \frac{x}{l}$

Example 5

A steel wire which was originally 2 m long is stretched by 3 mm. Find the strain.

Solution

Change in length = 3 mm = 0·003 m

original length = 2 m

$$\text{strain} = \frac{\text{change in length}}{\text{original length}}$$

$$= \frac{0{\cdot}003}{2}$$

$\therefore$ strain $= 0{\cdot}0015$ or $1{\cdot}5 \times 10^{-3}$

Example 6

The strain in a 5 m column is 2×10^{-4}. Find the change in length in mm.

Solution

Strain, $\epsilon = \frac{x}{l}$

$$\therefore x = \epsilon l$$

$$= 2 \times 10^{-4} \times 5$$

$$= 10 \times 10^{-4}$$

$$= 0{\cdot}001 \text{ m or } 1 \text{ mm}$$

$\therefore$ change in length of the column is 1 mm

11.4 Elasticity

When we dealt with testing a specimen of a material to destruction we were only interested in the load, the extension and the graph of F against x. We could easily have noted the length of the specimen and its cross-sectional area. This extra information would have enabled us to calculate the stresses caused by the increasing loads and also the resulting strains. Therefore instead of drawing a graph of load against extension we could have drawn a graph of stress against strain.

The shape of the graph would be exactly the same, the only difference is the scale used on each axis. Therefore Hooke's Law can be stated in terms of stress and strain as well as in terms of load and extension: the stress is directly proportional to the strain, provided it is within the elastic range for the material.

stress $\propto$ strain

or $\frac{\text{stress}}{\text{strain}}$ = a constant for the material

This constant is the slope of the straight line portion on the stress/strain graph and it is a measure of how elastic a material is. It is such an important constant for a material that it is called the **modulus of elasticity** and is given the symbol E. In some books it is called Young's modulus of elasticity.

Modulus of elasticity $= \dfrac{\text{stress}}{\text{strain}}$

or in symbols,

$$E = \frac{\sigma}{\epsilon}$$

The units of the modulus of elasticity are the same as the units of stress because strain has no units (remember that strain is a pure ratio). Therefore the modulus of elasticity is also measured in Pa, kPa, MPa or GPa.

The modulus of elasticity for a material is constant, no matter what shape or size. Some examples of E for different materials are listed below in GPa, i.e. $\times\ 10^9$ Pa.

Material	Modulus of elasticity (in GPa)
Steel	207
Copper	115
Brass	100
Aluminium	70

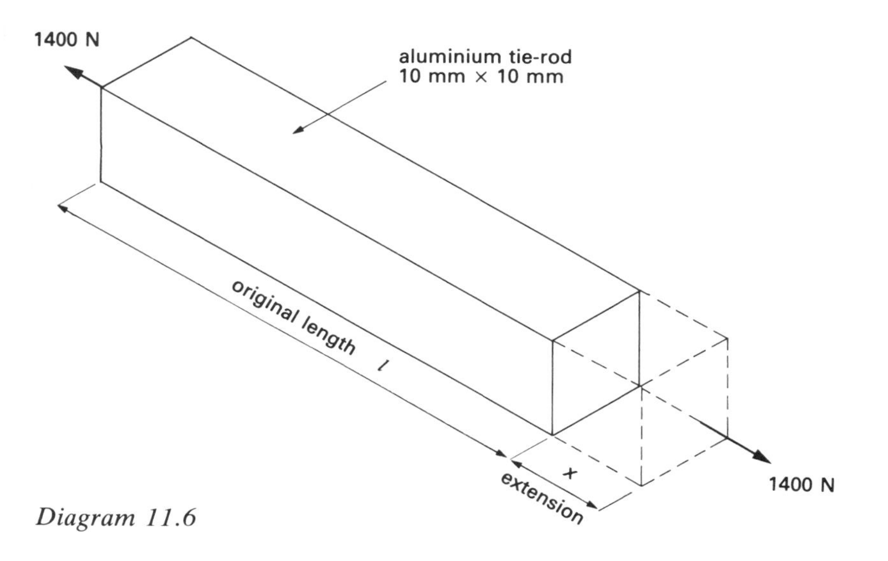

Diagram 11.6

Example 7

An aluminium tie-rod (Diagram 11.6) is 1·5 m long and its cross-section measures 10 mm by 10 mm. The tie-rod is resisting a tensile load of 1400 N and E for aluminium is 70×10^9 Pa. Find:

i the stress in the tie-rod;
ii the resulting strain;
iii the extension produced by this load.

Solution

i Stress $= \dfrac{\text{load}}{\text{cross-sectional area}}$

$= \dfrac{1400}{10 \times 10 \times 10^{-6}}$

∴ stress $= 14 \times 10^6$ Pa or 14 MPa

ii Modulus of elasticity, $E = \dfrac{\text{stress}}{\text{strain}}$

∴ strain $= \dfrac{\text{stress}}{E}$

$= \dfrac{14 \times 10^6}{70 \times 10^9}$

∴ strain $= 2 \times 10^{-4}$ or 0·0002

iii Strain $= \dfrac{\text{change in length}}{\text{original length}}$

∴ change in length = strain × original length

= 0·0002 × 1·5

= 0·0003 m

∴ change in length = 0·3 mm

Example 8

The support columns for a bridge (Diagram 11.7) are 12 m long and each has a cross-sectional area of 0·24 m². The stress must not be greater than 540 MPa. (Take E = 200 GPa.) For one column find:

i the maximum allowable load;
ii the maximum strain;
iii the maximum change in length.

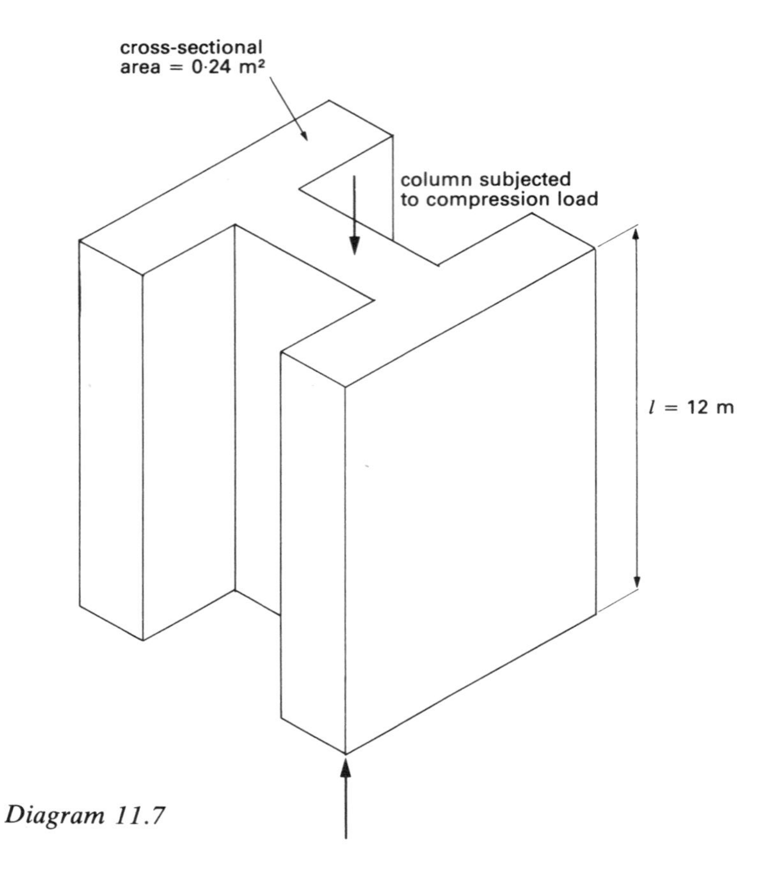

Diagram 11.7

Solution

i Maximum stress $= \dfrac{\text{maximum load}}{\text{cross-sectional area}}$

$\therefore 540 \times 10^6 = \dfrac{F}{0{\cdot}24}$

$\therefore F = 0{\cdot}24 \times 540 \times 10^6$

$= 130 \times 10^6$ N

$\therefore$ maximum load $= 130$ MN

ii To find maximum strain;

modulus of elasticity $= \dfrac{\text{stress}}{\text{strain}}$

$\therefore$ maximum strain $= \dfrac{\text{maximum stress}}{\text{modulus of elasticity}}$

$= \dfrac{540 \times 10^6}{200 \times 10^9}$

$\therefore$ maximum strain $= 2{\cdot}7 \times 10^{-3}$

iii Strain $= \dfrac{\text{change in length}}{\text{original length}}$

$\therefore$ change in length $=$ strain $\times$ original length

$= 2{\cdot}7 \times 10^{-3} \times 12$

$= 0{\cdot}0324$ m or $32{\cdot}4$ mm

$\therefore$ maximum change in length $= 32{\cdot}4$ mm

11.5 Factor of safety

When a structure or component is being designed, the designer must remember certain things. For example, very high quality materials are expensive; very accurate dimensions are expensive to produce; all kinds of loading must be considered (wind, expansion due to heat, etc.); the quality of bolts, rivets or welds may vary; the materials may be liable to corrode; there may be very high stresses during the construction; and so on.

The actual load which the structure or component is designed to carry is only one factor in a complicated process. Very often, designers estimate all the possible effects and finally decide on what will be a safe working load for the structure. This is worked out as a fraction of the maximum load at which the structure would be in danger of collapsing. (The maximum load corresponds to the highest load point on the load/extension graph for the material, i.e. beyond the elastic limit.)

Several modern structures are designed deliberately to operate very close to the maximum load point, e.g. the aircraft industry (because of weight).

In some text-books the maximum load is referred to as the ultimate load for a material; both are equally correct names. The **factor of safety** is the ratio of the maximum (ultimate) load to the safe working load (the load which can be allowed).

This can also apply to the stresses, because the load/extension graph is the same as the stress/strain graph. Therefore the factor of safety is also the ratio of the maximum (ultimate) stress to the safe working (allowable) stress.

$$\text{Factor of safety} = \frac{\text{maximum load}}{\text{safe working load}} = \frac{\text{maximum stress}}{\text{safe working stress}}$$

Example 9

The maximum load for a component is estimated to be 200 kN and the safe working load is to be 50 kN. What is the factor of safety for the component?

Solution

$$\text{Factor of safety} = \frac{\text{maximum load}}{\text{safe working load}}$$
$$= \frac{200 \times 10^3}{50 \times 10^3}$$
$$\therefore \text{factor of safety} = 4$$

Example 10

A material has an ultimate stress of $1{\cdot}2 \times 10^9$ Pa and a factor of safety of 5 is to be used. Find the allowable stress.

Solution

$$\text{Factor of safety} = \frac{\text{ultimate stress}}{\text{allowable stress}}$$
$$\therefore \text{allowable stress} = \frac{\text{ultimate stress}}{\text{factor of safety}}$$
$$= \frac{1{\cdot}2 \times 10^9}{5}$$
$$= 0{\cdot}24 \times 10^9 \text{ Pa}$$
$$\therefore \text{allowable stress} = 240 \times 10^6 \text{ Pa or } 240 \text{ MPa}$$

Example 11

The cross-sectional area of a strut is $0{\cdot}1$ m^2 and it is to carry working loads up to 21 MN. If the ultimate stress for the material is 630 MPa find the factor of safety.

Solution

$$\text{Allowable stress} = \frac{\text{load}}{\text{cross-sectional area}}$$
$$= \frac{21 \times 10^6}{0{\cdot}1}$$
$$= 210 \times 10^6 \text{ Pa}$$
$$\text{factor of safety} = \frac{\text{ultimate stress}}{\text{allowable stress}}$$
$$= \frac{630 \times 10^6}{210 \times 10^6}$$
$$\therefore \text{factor of safety} = 3$$

Example 12

A hoisting wire is made from a material for which the ultimate stress is 850 MPa and the stress at the elastic limit is 625 MPa. The wire has a working load of $2{\cdot}5$ kN. Find:

i the factor of safety to keep the stress just within the elastic limit for the material;
ii the diameter of the wire for this factor of safety;
iii the diameter of the wire for a factor of safety of 3.

Solution

i $$\text{Factor of safety} = \frac{\text{ultimate stress}}{\text{stress at elastic limit}}$$
$$= \frac{850 \times 10^6}{625 \times 10^6}$$
$$\therefore \text{factor of safety} = 1{\cdot}36$$

ii $$\text{Stress} = \frac{\text{load}}{\text{cross-sectional area}}$$
$$\therefore \text{cross-sectional area} = \frac{\text{load}}{\text{stress}}$$
$$= \frac{2{\cdot}5 \times 10^3}{625 \times 10^6}$$
$$= 4 \times 10^{-6} \text{ m}^2 \text{ or } 4 \text{ mm}^2$$
$$\text{But } \frac{\pi d^2}{4} = \text{area of circle}$$
$$\therefore \frac{\pi d^2}{4} = 4$$
$$\therefore d^2 = \frac{4 \times 4}{\pi}$$
$$= 5{\cdot}09$$
$$\therefore d = \sqrt{5{\cdot}09}$$
$$= 2{\cdot}26 \text{ mm}$$
$$\therefore \text{for a factor of safety of } 1{\cdot}36, d = 2{\cdot}26 \text{ mm}$$

iii For a factor of safety of 3,

$$\text{allowable stress} = \frac{\text{ultimate stress}}{\text{factor of safety}}$$

$$= \frac{850 \times 10^6}{3}$$

$$= 283 \times 10^6 \text{ Pa}$$

$$\text{cross-sectional area} = \frac{\text{load}}{\text{stress}}$$

$$= \frac{2{\cdot}5 \times 10^3}{283 \times 10^6}$$

$$= 8{\cdot}82 \times 10^{-6} \text{ m}^2 \text{ or } 8{\cdot}82 \text{ mm}^2$$

$$\therefore \frac{\pi d^2}{4} = 8{\cdot}82$$

$$\therefore d^2 = \frac{4 \times 8{\cdot}82}{\pi}$$

$$= 11{\cdot}2$$

$$\therefore d = \sqrt{11{\cdot}2}$$

$$= 3{\cdot}35 \text{ mm}$$

$\therefore$ for a factor or safety of 3, $d = 3{\cdot}35$ mm

Exercises

1 A material, being tested within its elastic range, extended $0{\cdot}8 \times 10^{-3}$ mm because of a load of 2000 N. What would its extension be for a load of 1500 N?

2 A load of 1·6 kN produced an extension of 0·002 m on a wire without causing any permanent damage. What load would produce an extension of 0·75 mm?

3 A stepped column, as shown in Diagram 11.8, is to carry a load of 180 kN. Find the stress in each part of the column.

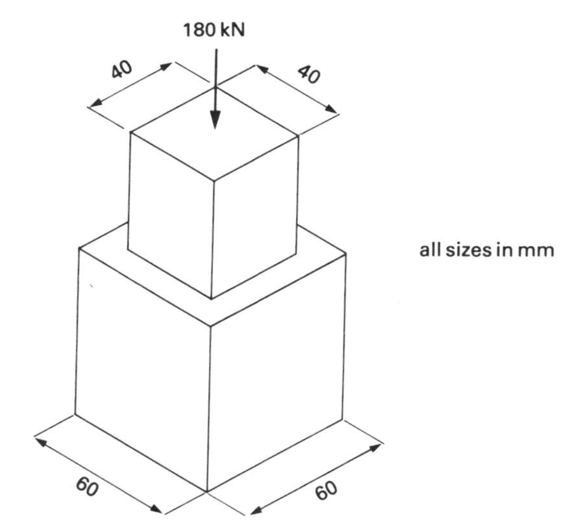

Diagram 11.8

4 The stress in a rod is not to exceed 65 MPa while carrying a tensile load of 1300 N. Find the necessary cross-sectional area of the rod in mm^2.

5 The diameter of an aluminium tie-wire is 1·2 mm and the allowable stress for the wire is 30 MPa. Find the allowable load for the wire.

6 A tow-bar, 3 m long, is compressed by 0·45 mm during braking. Find the strain.

7 The allowable strain on a bar is 0·002 and its length is 2·5 m. Find the allowable change in length.

8 During testing, a shaft was compressed by 0·06 mm. If the resulting strain was 75×10^{-6} what was the original length of the shaft?

9 During a tensile test on a specimen of copper it was found that the elastic limit was reached when the load was 690 N. The extension at this load was 0·06 mm. The gauge length of the specimen was 100 mm and the cross-sectional area measured 5 mm by 2 mm. Find: **i** the stress at the elastic limit, **ii** the strain at the elastic limit and **iii** the modulus of elasticity for copper.

10 The strain in a component was found to be $2{\cdot}7 \times 10^{-4}$. If E for the material is 110 GPa, find the stress.

11 A hydraulic car lift has a central column whose cross-sectional area is 2460 mm^2. The lift has to cope with vehicles weighing up to 14·5 kN. If E for the material of the column is 207×10^9 Pa, find the strain in the column at the highest load.

12 When a mass of 30 kg was suspended on the end of a wire 4 m long, it produced an extension of 6 mm. Find the cross-sectional area of the wire. (Take E = 100 GPa.)

13 A cast iron strut has a cross-sectional area of 7850 mm^2 and its modulus of elasticity is 55 GPa. If the strain is calculated to be 5×10^{-4}, find the force producing the strain.

14 The allowable stress in a trapeze wire is 60 MPa and the ultimate stress for the material is 420 MPa. Find the factor of safety.

15 The maximum stress for mild steel was estimated to be 1000 MPa. A screw-jack is to be made from this material and is to have a factor of safety of 4. The jack will have to cope with loads up to 6 kN. Find the minimum cross-sectional area for the screw of the jack.

Chapter 12

Heat and Energy Converters

12.1 Sources of thermal energy

Thermal energy (i.e. heat) is a very useful form of energy. It can be obtained by burning solids, liquids or gases, by collecting the radiation from the sun, by a nuclear reaction or by using the natural heat in the earth's crust.

Fossil fuels

These are fuels which are found in the earth. They were made from the remains of plants and animals millions of years ago. Heat and pressure caused these remains to form the substances we now use as fuels; coal, peat, oil and natural gas. There is more coal in the earth than any other fossil fuel but it is a very difficult and dangerous job getting the coal out. Peat is cut from the surface in layers in certain parts of Scotland and in Eire, but it contains a lot of moisture and does not produce as much heat as coal.

Crude Oil collects in pockets in the earth's crust. Finding new sources of oil in sufficient quantities is becoming more and more difficult. Oil companies have been exploring dangerous areas of sea for many years, looking for large pockets of oil under the sea-bed. The North Sea is one of the deepest and most hazardous areas and yet engineers have found ways of extracting that oil. Natural Gas was available in this country long before the oil reserves were discovered, but extracting the gas has similar problems.

BP's drilling rig 'Sea Conquest', east of the Shetland Islands

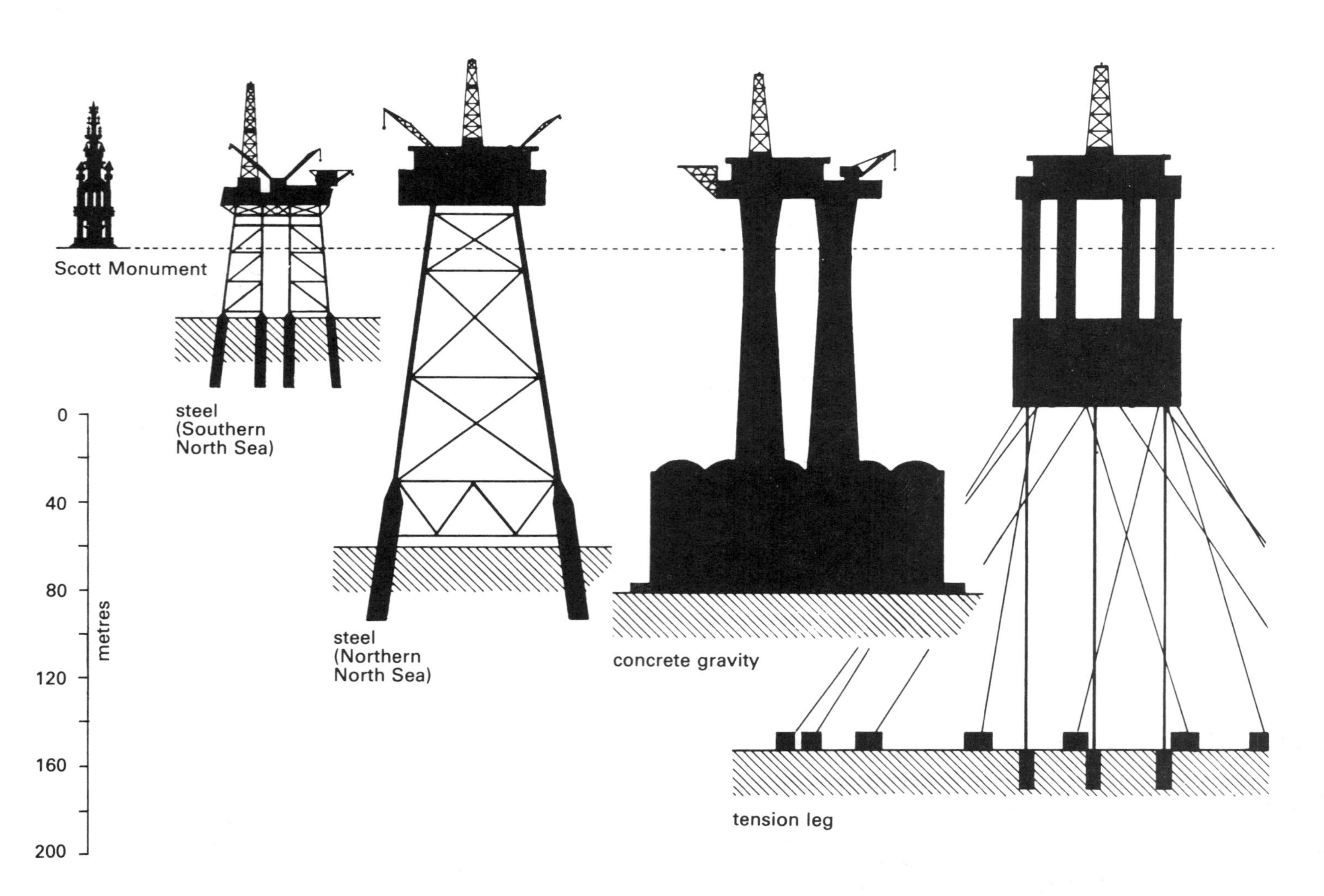

Comparative sizes of production platforms

In the not too distant future, the natural reserves of oil and gas will run out or at least become so small that it will be too expensive to extract them. Eventually, coal will become a major fuel again and engineers will be needed who will be able to solve the problems of mining coal from all kinds of awkward seams. Experts believe that there is still enough coal in the earth to last about 1500 years.

Solar energy

Energy from the sun travels through space as electromagnetic radi-

Houses fitted with solar panels at the Building Research Establishment

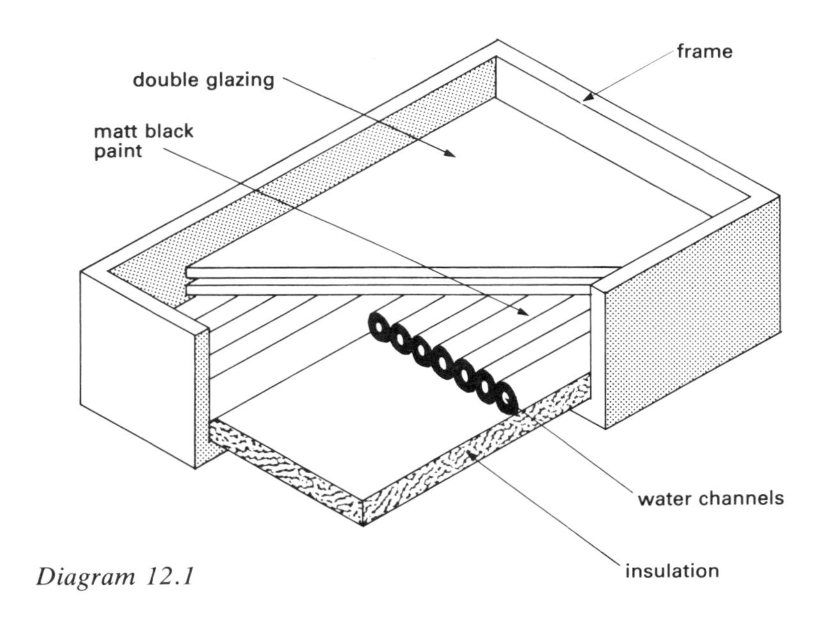

Diagram 12.1

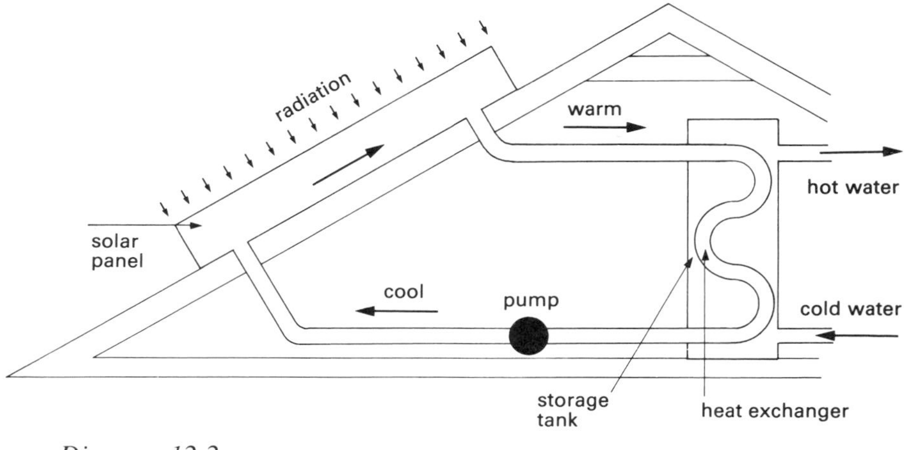

Diagram 12.2

Skylab in orbit with its wings of solar cells deployed

ation. Flat-plate collectors are simple devices which are used throughout the world for small jobs like heating water and buildings. Such collectors, as shown in Diagram 12.1, are often called **solar panels**.

Solar panels work during daylight hours and not only in direct sunlight. The matt black plate is warmed by the sun's radiation and it heats up. This heat is then collected by the water which flows along the channels or pipes on the back of the plate. To improve the efficiency of a panel, it can be insulated and double glazed. The warmed water from a solar panel is circulated through a heat exchanger in the storage tank as shown in Diagram 12.2.

Another way of using the sun's radiation is to focus the rays onto a small area. In some parts of the world, very large parabolic reflectors have been built and the concentrated radiation from the sun is used as a **solar furnace** for boilers, etc. Simple solar furnaces can create temperatures as high as 2700°C, but they need direct sunlight.

Solar Cells are small devices which convert the sun's radiation into electricity. However, they are so expensive and inefficient (about 10%) that they are not likely to be of importance other than as a reliable source of electrical energy for satellites.

Nuclear energy

At present this provides only about 2% of the energy used in the world. The energy comes from the splitting of atoms, usually of uranium. This is carefully controlled in a nuclear reactor, but there are many problems connected with nuclear power, mainly leakage and storage of radioactive waste. The advantage of using nuclear power is that although very little fuel is used, a large amount of energy is released. (1 kg of uranium can produce as much useful energy output as 8 tonnes of coal.)

Geothermal energy

This is available in certain parts of the world where hot water or steam flows out of the earth. When water comes in contact with hot, underground rocks it becomes heated and can turn into steam. Commercial companies search for areas where steam is trapped underground so that they can tap the steam and then build a steam turbine

Wylfa Nuclear power station

generating station close by. Sometimes, when there is no steam, but the underground rocks are hot, water can be pumped down to the hot rocks to create the supply of steam. Geothermal power plants are very clean. They do not produce smoke and do not produce poisonous waste products. Several countries use such power plants, e.g. U.S.A., New Zealand, Italy and Japan.

12.2 Temperature and heat energy

Temperature is a measure of the degree of hotness (or coldness) of a body. Heat energy is the amount of energy contained by the molecules of a body. It is the kinetic energy of the molecules. The relationship between heat and temperature can be shown in a general way.

Consider two beakers of water, as shown in Diagram 12.3, which are heated by identical bunsen burners. Both burners are supplying identical amounts of heat energy but it is fairly obvious that the water in B will get hotter more quickly. Therefore, for the same amount of heat energy, the smaller mass (B) has to store more energy per molecule than the bigger mass (A). This energy is stored by the molecules as kinetic energy, i.e. the molecules move at higher speeds. The level of activity of the molecules is measured by the thermometer. Therefore the thermometer in B will show a higher temperature than the one in A even though the heat energy in both beakers is the same.

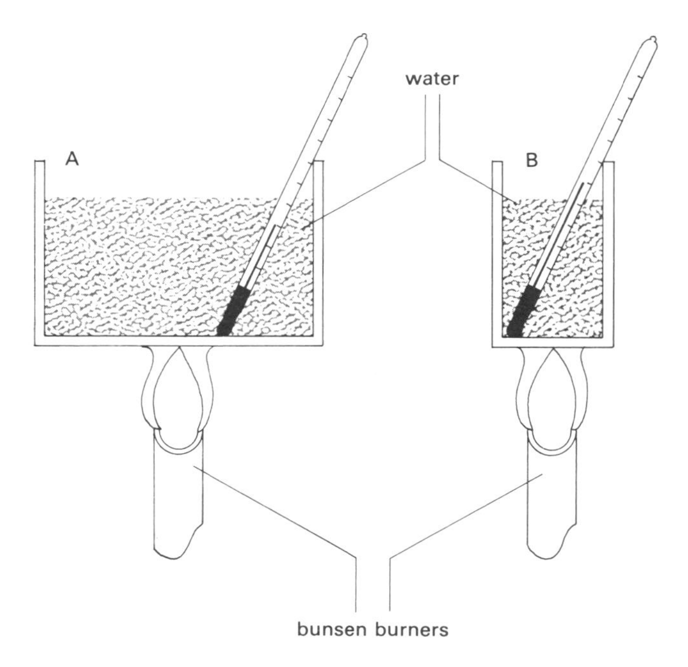

Diagram 12.3

Temperature scales

Different levels of temperature are measured on a scale. In SI Units, the scale used is called the **Celsius** scale. Zero of this scale is the temperature when water just starts to freeze, and the temperature when water just starts to boil is called 100 degrees Celsius (or 100°C). Between the freezing point of water and the boiling point of water there are 100 equal divisions called degrees.

The zero for the Celsius scale is really a random point to start measuring from. Temperatures can, and do, go below zero. Scientists have discovered, however, that there is a limit to how cold any matter can go. That limit is called the absolute zero of temperature; nothing can get colder than that. The absolute zero is at −273°C.

A new temperature scale can be worked out now, using the absolute zero as the zero of the scale, and the temperatures will therefore be absolute temperatures. This new scale is called the **kelvin** scale. The size of the intervals on this scale are exactly the same as the intervals on the Celsius scale.

1 degree interval on Celsius = 1 degree interval on kelvin

(The unit symbol for degree Celsius is °C and for degree kelvin it is K.) We can see the relationship between the two scales as follows:

boiling point of water	= 373 K or	100°C
freezing point of water	= 273 K or	0°C
absolute zero	= 0 K or	−273°C

Therefore: absolute temperature (in degrees kelvin) = common temperature (in degrees Celsius) + 273

When using symbols for temperature we must be able to tell the difference between a common temperature measured in degrees Celsius and an absolute temperature measured in degrees kelvin. Therefore for common temperature we will use the symbol t, and for absolute temperature we will use the symbol T.

i.e. $T = t + 273$

It has been mentioned that there is a limit to the lower end of a temperature scale. However, there does not seem to be any upper limit; temperature scales have no maximum. The centre of the sun is estimated to have a temperature of about 15 000 000°C.

Thermometers

There are three common types of thermometer; the liquid-in-glass, the bi-metallic and the electrical.

1 Liquid-in-glass thermometers use the principle that liquids expand with heat. These are very common thermometers and can vary in size. They usually have either mercury or coloured alcohol as the liquid and are not suitable for very high temperatures.

2 Bi-metallic strip thermometers use the principle that most solid materials expand with heat. A simplified arrangement is shown in Diagram 12.4. As the strip becomes warmer, the brass and steel expand together, but, because brass expands more than steel does, the end of the coiled bi-metallic strip moves clockwise. This principle is often used in thermostats (heat switches) for controlling temperature. As the strip moves it switches the power on and off.

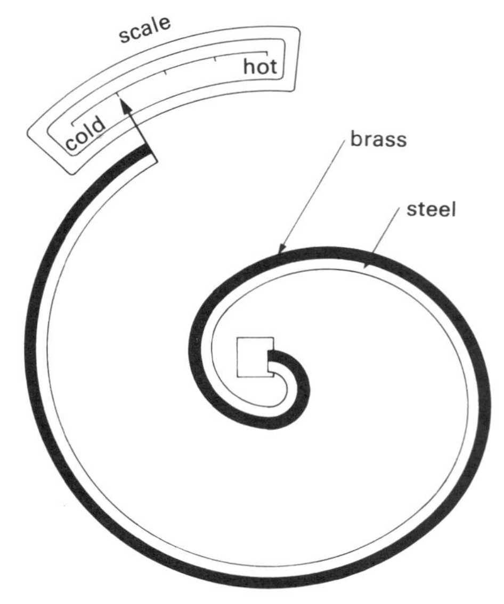

Diagram 12.4 coiled bi-metallic strip

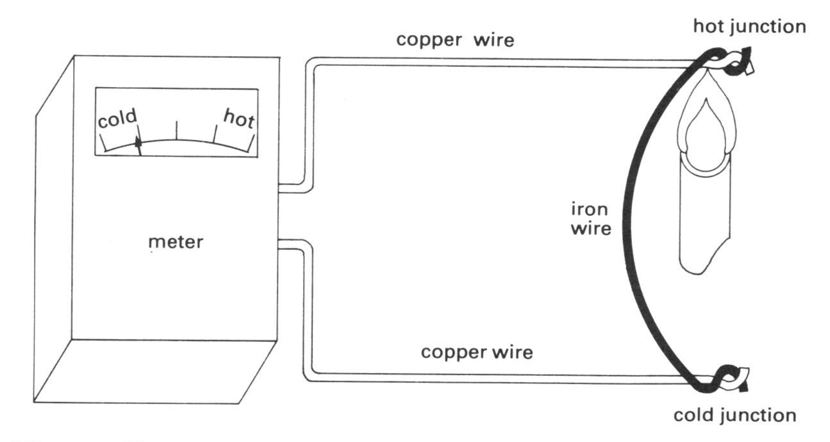

Diagram 12.5

3 Electric Thermometers (Thermocouples) use the principle that an electromotive force is created at the junction of two different metals. As the temperature of the junction changes, the e.m.f. changes. This change in e.m.f. can be recorded on a sensitive electric meter as illustrated in Diagram 12.5.

The scale on the meter is usually marked in degrees to give a direct temperature reading. The reading is the temperature difference between the hot and cold junctions. This kind of thermometer, with other metals, is useful for measuring very high temperatures (up to 1500°C) or where temperatures have to be recorded some distance away. For greater accuracy, the cold junction should be kept at a regulated temperature.

12.3 Heat transfer

Heat behaves rather like water or electricity because it flows from high pressure to low pressure. For heat, the pressure is the level of activity of the molecules, i.e. the temperature. Therefore heat flows from a body at a high temperature to a body at a low temperature. There are three main ways in which the heat can be transferred.

Conduction – where heat is passed from molecule to molecule in a body, rather like conducting electricity.

Convection – where the heat is transferred by a mass movement of molecules, like circulating air or water.

Radiation – where the heat is transmitted across space (even across a vacuum) in the same way as the sun's energy.

Usually all three ways of transferring heat happen together, but one of the ways may be transferring most of the heat, e.g. in a radiant fire, radiation is most important. In a convector heater convection currents are most important, and in a soldering iron, conduction is most important, etc.

Energy is so precious that we must not waste it by allowing it to pass into the atmosphere. We are encouraged to conserve heat energy by reducing the amount of heat flowing out of our homes. Heat energy escapes from buildings in many ways; through the loft, walls, doors, windows and also from hot water tanks and pipes. This heat loss can never be completely stopped, but it can be reduced by attending to each area in turn. Some areas of heat loss can be dealt with fairly easily and cheaply but others can be very expensive.

Methods of reducing heat loss

1 Draughtproofing Although a supply of fresh air is essential, excessive draughts from badly fitting windows and doors must be stopped. As cold air comes into the building, costly warm air goes out. Draughtproofing is quite cheap.

2 Loft insulation We know that warm air rises, therefore it is sensible to insulate the loft to stop too much heat being lost through the ceiling. Usually 80 mm thickness of insulation is used, but a 100 mm thickness of insulation can be used over very warm rooms. Making the layers thicker than this may not be worth the expense. The materials for this are usually of fibre-glass blankets or loose pellets of polystyrene. Loft insulation is considered to be one of the cheapest and most effective first steps in insulating a building.

3 Cavity wall insulation Most houses are now built with a double outer wall. The air-space between the inner and outer bricks is called the wall cavity. It is quite common now-a-days for this cavity to be filled with a cellular plastic (a honeycomb of bubbles of air) as an insulating layer within the wall. In established houses, the plastic is injected into the cavity through holes drilled in the outer bricks. The plastic enters as a foam which fills the cavity completely and then hardens. Other materials can be used, for example fine strands of glass-fibre which are blown into the cavity to fill it. It is quite common for people to insulate their cavity walls after they have completed the first two stages of home insulation described above. It is a little more expensive than good loft insulation but is still considered to be worthwhile.

4 Double glazing A single pane of glass allows a lot of heat to escape through the glass by conduction and radiation. This heat loss can be considerably reduced by putting two panes of glass into the same window frame. The air gap between the two panes is quite important. In practice, the best size of gap for reducing heat loss is about 12 mm. Some companies manufacture sealed units which are two panes of glass 2 to 3 mm apart, sealed at their edges after the air in the gap has been extracted.

There are many possible arrangements for double-glazing, some fairly expensive and some very expensive, but none are very cheap. Whether it is worthwhile or not really depends on how big the windows are in relation to the walls.

Double glazing of windows or glass doors is usually the last stage of domestic insulation.

5 Water pipes Insulating water pipes and water tanks really goes with loft insulation (item 2). By insulating the loft it will become colder and therefore there would be an even greater heat loss from the hot water tank and hot water pipes if they are in the loft. Also the cold water tank and cold pipes would become even colder in winter, with the danger of freezing and causing bursts. Tanks are usually wrapped in a jacket of insulating material, e.g. glass-fibre, and pipes are lagged with either strips of insulating material or pre-formed cellular plastics. Insulating tanks and water pipes is fairly cheap and should be done at the same time as loft insulation.

Storing thermal energy

Off-peak electric storage heaters have already been mentioned in our course. They are one example of how heat is taken in, stored for a period and then allowed to pass to the surroundings.

When the off-peak heaters are being supplied with electrical energy, about half of this energy is given off to the surroundings immediately as direct heat. The other half of the energy is stored for use when the electricity supply is not available, i.e. at peak times. In order to store enough heat, the storage heater must contain a suitable material, one which can take in a lot of thermal energy. Such materials are usually fairly dense and the most commonly used materials for this are brick and concrete. Underfloor heating works on exactly the same principles.

Water is also used as a store for thermal energy where the heat has to be transported from the source of energy (the boiler) to the place where the heat is required (the radiators in the rooms). Therefore water stores and transports thermal energy in central heating systems.

12.4 Specific heat capacity

We have discussed heat, heat transfer, insulation and storage without actually measuring quantities of heat or quantities of heat transferred. It has been mentioned that some materials are good at storing heat but without saying why.

We will now look at how to work out quantities of heat transfer and to do this, we need to use another property of materials, the **specific heat capacity**.

The specific heat capacity of a material is the amount of heat energy that 1 kg of the material will take in or give out for a change in its temperature of 1 degree. The symbol for specific heat capacity is c and the units are joules per kilogram per degree kelvin, written as J/kgK.

Consider a mass of 1 kg which is heated until its temperature changes by 1 degree kelvin, then:

heat added = specific heat capacity

For heat energy, which is measured in joules, we use the symbol Q, therefore:

$Q = c$ for 1 kg and 1 degree kelvin

If the mass is m kg but the temperature change is still 1 degree, then:

$Q = mc$ for m kg and 1 K

If the temperature changes from T_1 to T_2, the number of degrees of temperature change is $(T_2 - T_1)$ which can be written as ΔT, and if we have a mass of m kg, then,

$$Q = mc\ (T_2 - T_1)$$

or $Q = mc\,\Delta T$

This is called the **heat transfer equation** and in words the heat transfer is equal to the mass times the specific heat capacity times the number of degrees of temperature change. Different materials have different values of specific heat capacity.

Material	Specific heat capacity (c) ($\times 10^3$ J/kg K)
Water	4·2
Paraffin oil	2·2
Firebrick	0·84
Glass	0·67
Iron	0·46
Copper or zinc	0·39

Example 1

How much heat energy is required to raise the temperature of 1·5 kg of water from 6°C to 100°C? (Take the specific heat capacity of water to be $4{\cdot}2 \times 10^3$ J/kg K.)

Solution

Mass, $m = 1{\cdot}5$ kg
specific heat capacity, $c = 4{\cdot}2 \times 10^3$ J/kg K
temperature change, $\Delta T = (100 - 6) = 94$ K degrees
heat transfer, $Q = mc\,\Delta T$
$= 1{\cdot}5 \times 4{\cdot}2 \times 10^3 \times 94$
$= 592 \times 10^3$ J
$\therefore$ heat energy required = 592 kJ

Example 2

An immersion heater raises the temperature of 8 kg of water from 4°C to 68°C in 21 minutes. Find the power rating of the heater, assuming it to be 100% efficient.

Solution

Heat transfer = mass × specific heat capacity × temperature change
$\therefore Q = mc\,\Delta T$
$= 8 \times 4{\cdot}2 \times 10^3 \times (68 - 4)$
$= 2{\cdot}15 \times 10^6$ J

power $= \dfrac{\text{energy transfer}}{\text{time taken}}$
$= \dfrac{2{\cdot}15 \times 10^6}{21 \times 60}$
$= 1{\cdot}71 \times 10^3$ W
$\therefore$ power rating = 1·71 kW

Example 3

A 0·5 kg block of copper is heated in an oven to a temperature of 85°C. It is then dropped into a beaker of water which was at a temperature of 5°C. After a few minutes the temperature steadied at 10°C. Find the mass of water in the beaker, assuming no energy losses ($c_{\text{water}} = 4{\cdot}2$ kJ/kg K and $c_{\text{copper}} = 0{\cdot}39$ kJ/kg K).

Solution

Heat taken in by water = heat given out by copper
$\therefore (mc\,\Delta T)$ for water $= (mc\,\Delta T)$ for copper
$\therefore m \times 4{\cdot}2 \times 10^3 \times (10 - 5) = 0{\cdot}5 \times 0{\cdot}39 \times 10^3 \times (85 - 10)$
$\therefore m \times 21 \times 10^3 = 14{\cdot}6 \times 10^3$
$\therefore m = \dfrac{14{\cdot}6 \times 10^3}{21 \times 10^3}$
$= 0{\cdot}7$ kg
$\therefore$ mass of water in beaker = 0·7 kg

Example 4

A bath containing 15 kg of water at 10°C has 5 kg of boiling water added to it. Find the final temperature of the bath water.

Solution

Let the final temperature be t.

Heat given out by hot water = heat taken in by cold water
$\therefore (mc\,\Delta T)$ for hot water $= (mc\,\Delta T)$ for cold water
$\therefore 5 \times 4{\cdot}2 \times 10^3 \times (100 - t) = 15 \times 4{\cdot}2 \times 10^3 \times (t - 10)$

But $(4{\cdot}2 \times 10^3)$ is a common factor on both sides of the equation.

$\therefore 5(100 - t) = 15(t - 10)$
$\therefore 500 - 5t = 15t - 150$
$\therefore 20t = 650$
$\therefore t = 32{\cdot}5$°C
$\therefore$ common temperature, $t = 32{\cdot}5$°C

12.5 Energy converters

An engine is a device which changes energy from one form into another, usually from thermal energy to mechanical energy. Engines are of two types – reciprocating engines or turbines. The most common reciprocating engine is the ordinary car engine. It is called a reciprocating engine because working parts in the engine slide up and down, i.e. they reciprocate.

The energy is supplied to the engine as a fuel (chemical energy). This is burned and produces thermal energy which makes the gases expand and push the piston down. The piston in turn makes the crankshaft rotate, thus producing mechanical energy (Diagram 12.6).

Chemical energy → thermal energy → mechanical energy

There are two main types of reciprocating engine in everyday use. These are the petrol engine and the diesel engine. The petrol engine takes its fuel as a vapour, i.e. a mixture of petrol and air. The mixing takes place in the carburettor and the fuel vapour is then passed through the inlet valve into the cylinder where it is burned. To make the fuel start to burn, a spark is used which is supplied by the spark plug.

The diesel engine is slightly different. It takes air into the cylinder and when it has been compressed and is very hot, diesel fuel is forced into the cylinder through the fuel injector. Because of the high temperature, the diesel fuel starts burning spontaneously.

The petrol engine uses a carburettor to mix the correct amounts of air and petrol and the diesel engine uses a fuel injection pump to produce the high pressure needed to force the fuel into the cylinder. Both engines have inlet and exhaust valves which must open and close at exactly the right time and the fuel in the cylinder must burn at exactly the right time.

Turbines are engines which have rotating blades; they do not have reciprocating parts. There are a great number of special kinds of turbine, far too many to mention here, so we will only look at turbines in general terms. There are two different types of turbine which are designed to work in different ways.

One type of turbine depends on the force with which the fluid strikes its rotor (i.e. its rotating blades). The other type of turbine depends on the pressure of the fluid as it flows through its blades.

Different fluids can be used to drive turbines, e.g. water, steam, gas etc. Water turbines are normally slow-speed, but powerful. They are

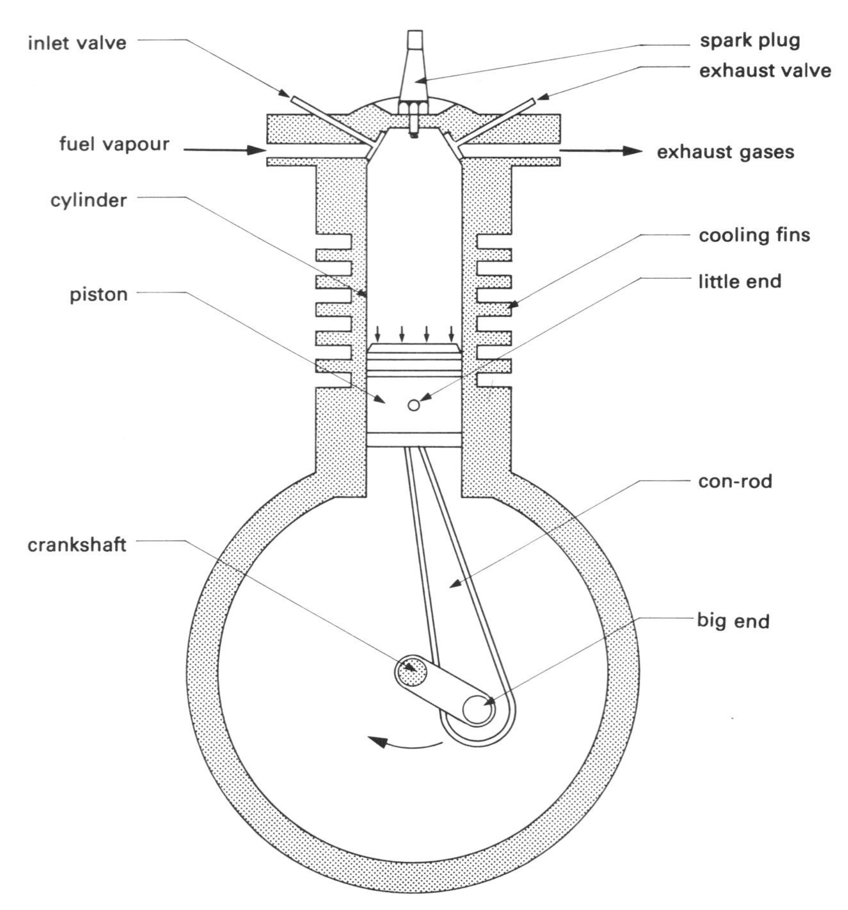

Diagram 12.6

Section of a Princess 2000 petrol engine and gear box

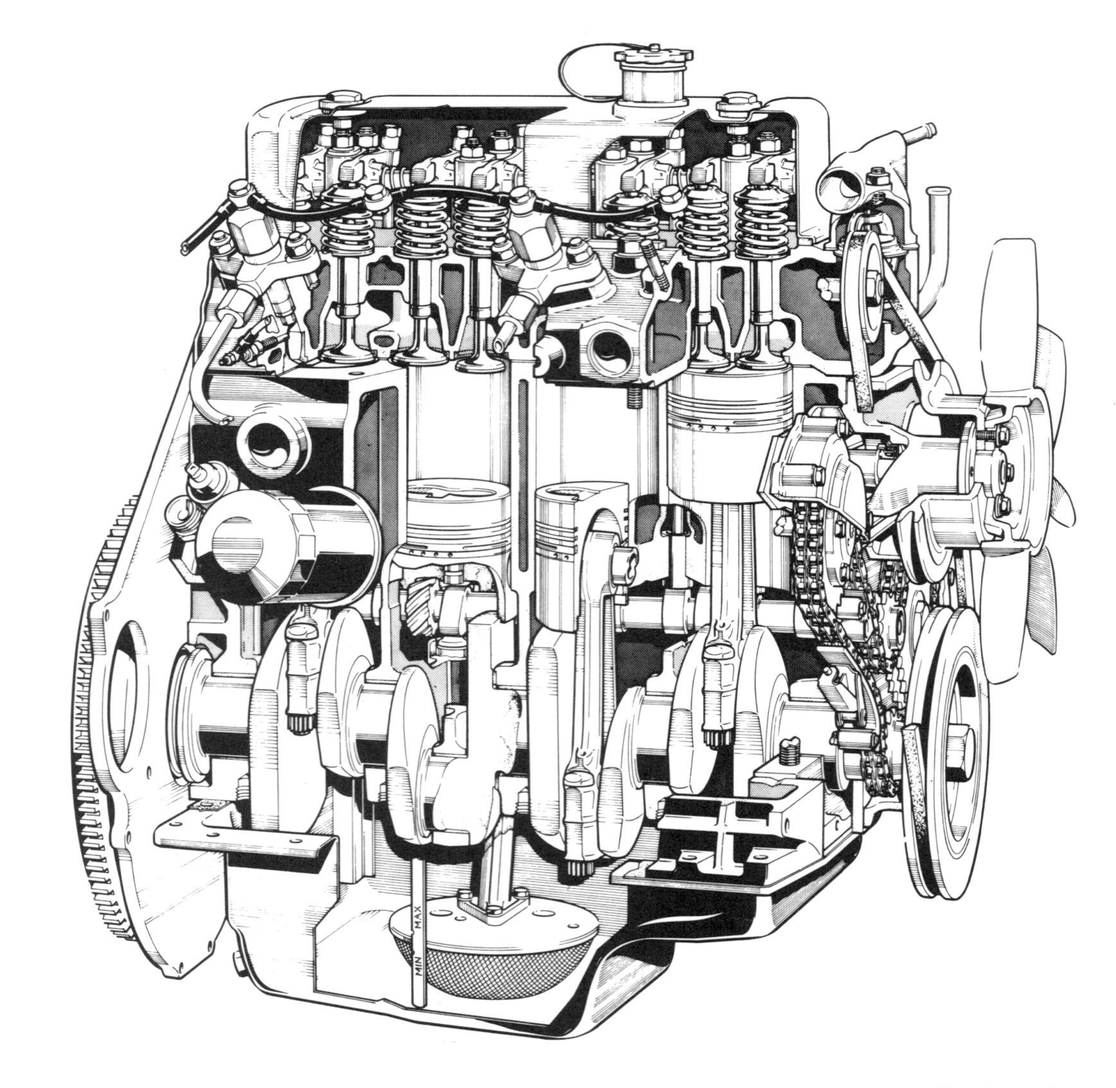

Section of an Austin Morris diesel engine

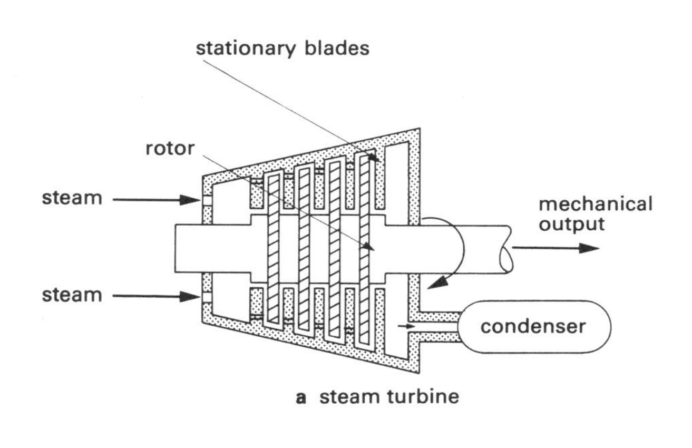

a steam turbine

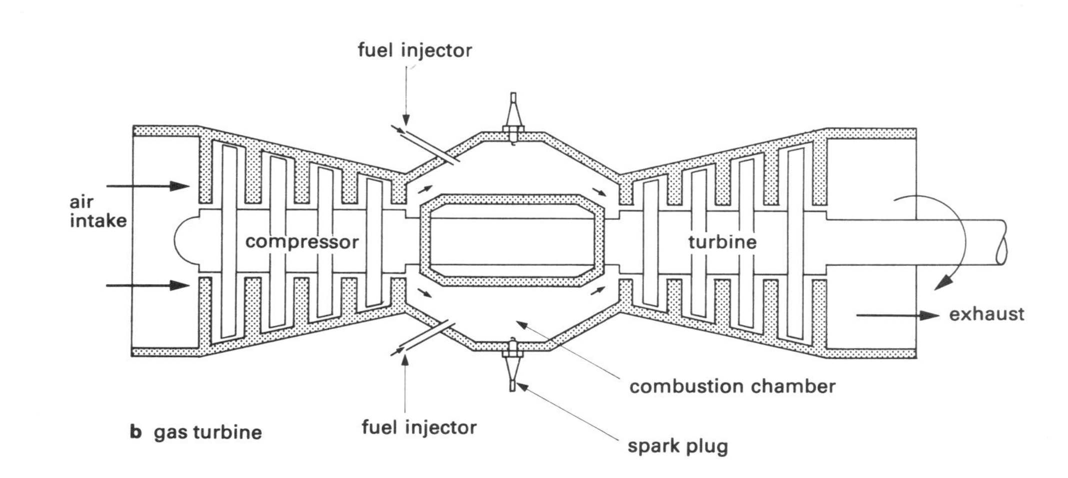

b gas turbine

Diagram 12.7

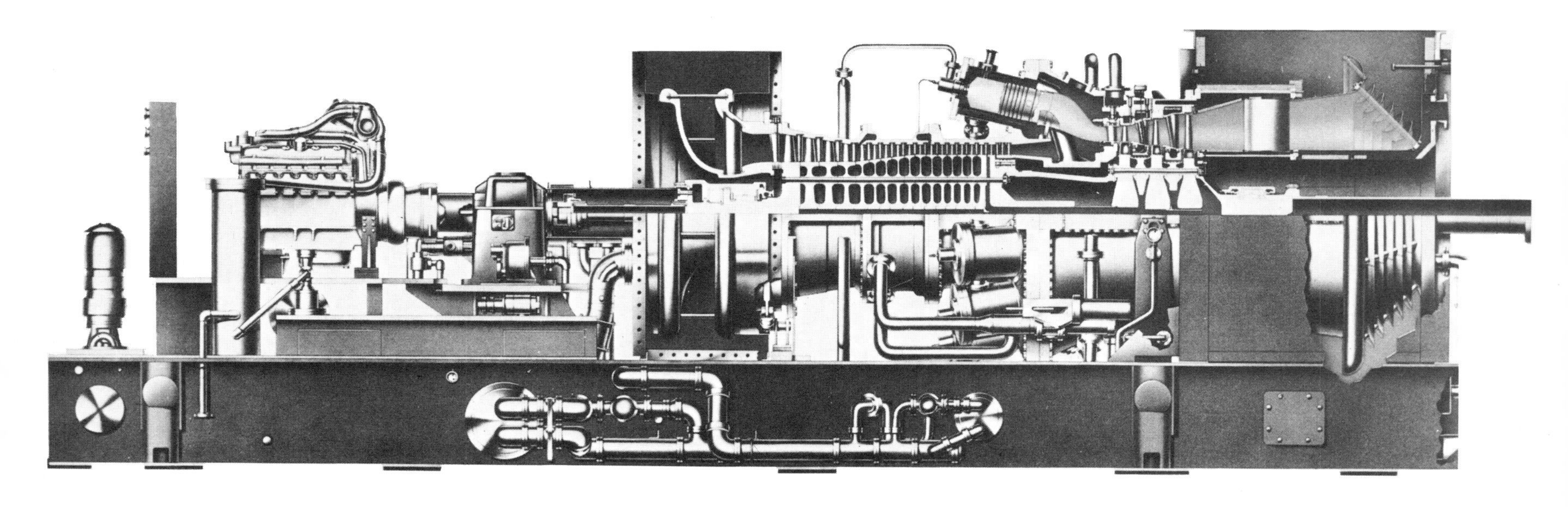

Section of a gas turbine engine

popular for hydroelectric power plants and are very large engines. Steam and gas turbines tend to be smaller in size and can have very high shaft speeds. Simplified arrangements for steam and gas turbines are shown in Diagram 12.7**a** and **b**. The function of the stationary blades is to direct the flow of the fluid onto the next set of moving blades on the rotor.

12.6 The 4-stroke cycle

Turbines operate in continuous flow, but reciprocating engines do not. They must operate in a set sequence, called a cycle of operations. The fuel must be let into the combustion chamber, trapped, burned and then pushed out. Thereafter the cycle repeats itself.

The 4-stroke cycle is the most common for petrol and diesel engines. When a piston moves from one end of the cylinder to the other, it is said to have completed 1 stroke. A piston which slides up and then down a cylinder has completed 2 strokes.

Therefore a 4-stroke cycle is one in which the piston moves up and down the cylinder twice. This corresponds to two revolutions of the crankshaft and the function of each stroke in the cycle is illustrated in Diagram 12.8.

The strokes

1 Induction The inlet valve opens while the piston travels down and fuel vapour is sucked into the cylinder. The exhaust valve is closed.

2 Compression With both valves closed, the piston moves up the cylinder compressing the fuel mixture into a small space, called the combustion chamber.

3 Power At top-dead-centre, or close to that, a spark from the spark plug sets fire to the compressed mixture. Since both valves are still closed, the hot expanding gases push the piston down with great force.

4 Exhaust As the piston starts to slide up the cylinder again, the exhaust valve opens. By the time the piston reaches the top of its stroke, the spent gases have been pushed out and the whole sequence is ready to start again.

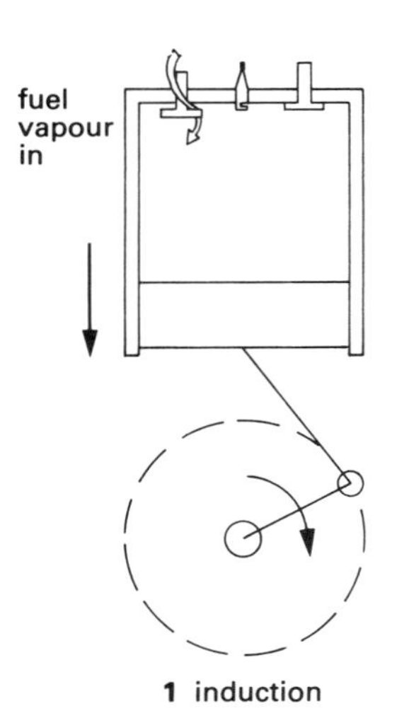

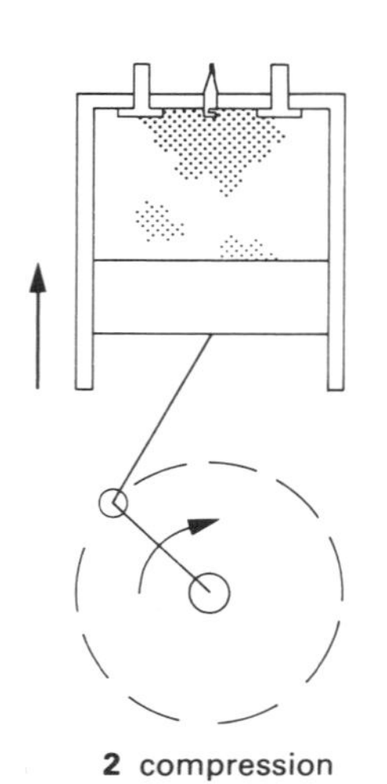

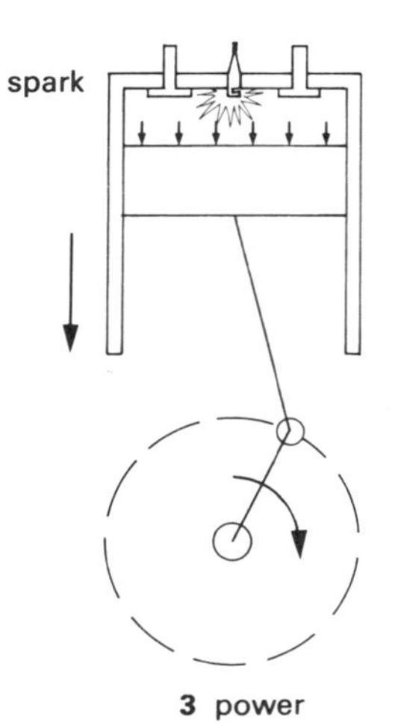

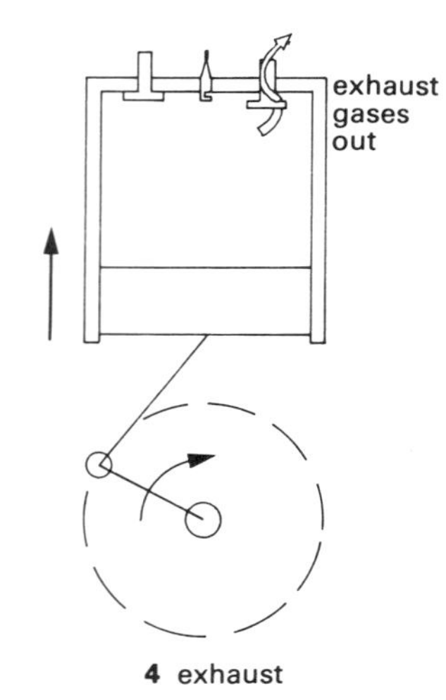

Diagram 12.8 4-stroke cycle

12.7 Specific energy of fuels

We have already seen that the chemical energy of a fuel is measured by the amount of thermal energy released when 1 kg (or 1 litre) of the fuel is burned (see Chapter 7). For example:

petrol = 30 to 36 MJ/kg (or 42 to 48 MJ/l)
diesel = 34 to 36 MJ/kg (or 40 to 42 MJ/l)

If we know the amount of fuel consumed then we can work out the thermal energy supplied. For example, an engine which consumed 5 litres of fuel of specific energy 45 MJ/l, has been supplied with 225 MJ of thermal energy.

12.8 Efficiency of energy conversions

There are many ways of working out how much of the energy input has been converted into a useful energy output. Calculating the thermal energy input is quite straightforward as we have already seen, but to find out how much mechanical energy is available as a useful output we must carry out some kind of test.

When testing an engine, it is usual to time the test period. This means that we can calculate input and output power. There are two simple tests which can be carried out to find the output power of an engine. These tests use our knowledge of torque and speed to find power. Both tests are like brakes on the output shaft from the engine, and for this reason the output power is sometimes called the **brake power** (P_b) of the engine.

From Chapter 7, we know that power is found from, $P = 2\pi nT$, where n is the rotational speed in rev/s and T is the torque on the shaft in N m. Therefore in each test we will need to find n and T in order to calculate P_b.

Prony brake

This is normally used for low speeds and it consists of a clamp which has friction pads on its inner surfaces. The device is lightly clamped onto the output shaft as shown in Diagram 12.9. Since the shaft is turning, it tends to drag the brake with it. The function of the load, F, on the load-arm is to balance this tendency. F must be just big enough to keep the load-arm horizontal during the test at constant speed. The speed of the shaft can be found by using a rev-counter.

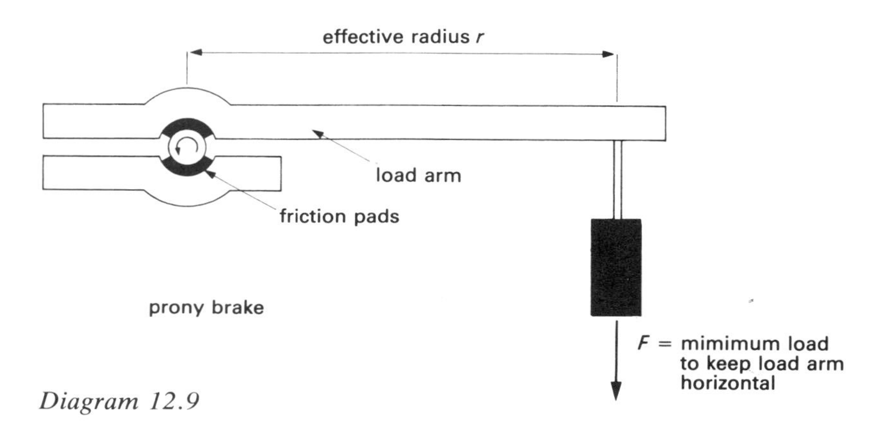

Diagram 12.9

From this information, we can calculate the output power (i.e. brake power, P_b) as follows:

$$P_b = 2\pi nFr$$

where n = shaft speed of engine in rev/s
F = load at end of load-arm in N
r = effective radius of load-arm in m

Example 5

The following information was obtained during a test on an engine using a prony brake. Shaft speed = 360 rev/min; effective length of load arm = 1·2 m; load = 15 N. Find the brake power of the engine.

Solution

$$P_b = 2\pi nFr$$
$$= \frac{2\pi \times 360 \times 15 \times 1{\cdot}2}{60}$$
$$= 679 \text{ W}$$

$\therefore$ brake power of engine = 0·679 kW

Rope brake

In this form of test, a rope is wound once round the flywheel of the engine. One end of the rope is attached to a spring balance and the other has a mass hung on it, as shown in Diagram 12.10. Note that the rotation of the flywheel is so as to try to lift the dead load.

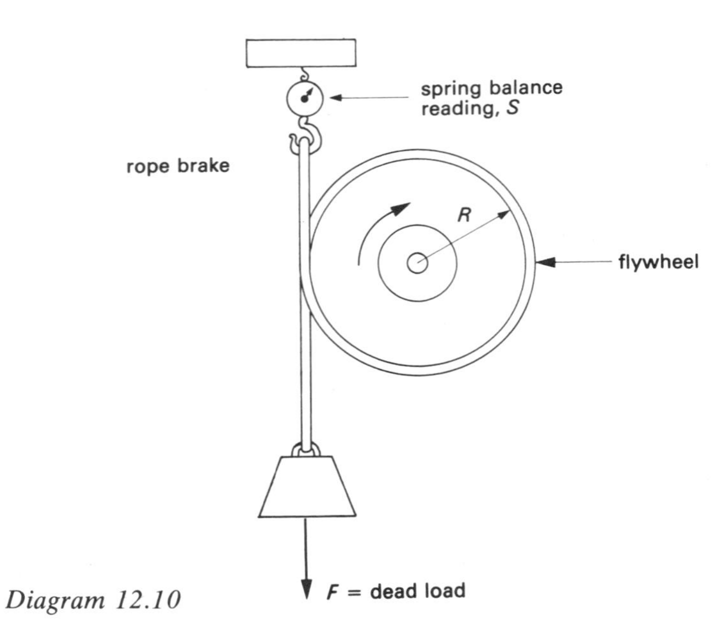

Diagram 12.10

During the test, when the flywheel is turning, the reading on the spring balance, S will be less than F, the dead load. The difference between these two quantities is the effective braking load.

$\therefore$ effective braking load $= F - S$ (newtons)

The thickness of the rope in these tests affects the radius at which the load acts.

$\therefore$ effective radius = flywheel radius + rope radius
$= R + r$ (metres)

But we know that torque = force × radius

$\therefore$ braking torque = effective load × effective radius
or $T_b = (F - S) \times (R + r)$

When dealing with rope brakes, it is best to work out the braking torque, T_b, before going on to calculate the brake power.

$P_b = 2\pi n T_b$ (where T_b = braking torque as already found)

Example 6

The following information was obtained during a test on an engine using a rope brake. Radius of flywheel, $R = 300$ mm; radius of rope, $r = 10$ mm; spring balance reading, $S = 35$ N; dead load, $F = 215$ N; speed of engine, $n = 30$ rev/s. Find the brake power of the engine under these conditions.

Solution

Braking torque, $T_b = (F - S)(R + r)$
$= (215 - 35)(0{\cdot}3 + 0{\cdot}01)$
$= 180 \times 0{\cdot}31$
$= 55{\cdot}8$ Nm

brake power, $P_b = 2\pi n T_b$
$= 2\pi \times 30 \times 55{\cdot}8$
$= 10{\cdot}5 \times 10^3$ W

$\therefore$ brake power of engine $= 10{\cdot}5$ kW

Brake thermal efficiency (η_{bth})

When we compare an output to an input we get the efficiency of the process. If both input and output are mechanical forms of energy, then we get a mechanical efficiency. If the input is thermal energy and the output is mechanical energy, then we get a thermal efficiency. If the input is thermal energy and the output is energy at a brake, then we get what is called **brake thermal efficiency**. The brake thermal efficiency of an engine is the ratio of the brake power output of the engine to the thermal power input to the engine.

$$\eta_{bth} = \frac{P_b}{\text{fuel power input}}$$

Example 7

Find the brake thermal efficiency of an engine which has a brake power output of 25 kW when consuming 10 litres of fuel per hour. The fuel has a specific energy of 31·5 MJ/l.

Solution

$P_b = 25 \times 10^3$ W

and fuel power input $= 31{\cdot}5 \times 10^6 \times \frac{10}{3600}$
$= 87{\cdot}5 \times 10^3$ W

$$\therefore \text{brake thermal efficiency} = \frac{25 \times 10^3}{87{\cdot}5 \times 10^3} = 0{\cdot}286 \text{ or } 28{\cdot}6\%$$

Efficiency of other energy converters

1 Electric motors can be tested to find the brake power in the same way as heat engines. However, the input power is electrical. Therefore, for electric motors,

$$\text{efficiency} = \frac{\text{brake power output}}{\text{electrical power input}}$$

$$\eta = \frac{P_b}{VI}$$

2 Electric kettles and heaters have an electrical energy input and a thermal energy output. The output thermal energy is the useful heat transfer to the fluid being heated.

$$\text{Efficiency} = \frac{\text{thermal energy output}}{\text{electrical energy input}}$$

$$\eta = \frac{mc\,\Delta T}{VIt}$$

3 Since steam and gas turbines are heat engines, their efficiency will be the brake thermal efficiency as already described. For a water turbine, the input energy is due to the potential energy of the reservoir.

$$\text{Efficiency} = \frac{\text{brake power output}}{\text{potential energy input per second}} = \frac{P_b}{\dot{m}gh}$$

where $\dot{m}$ is the mass flow rate in kg/s.

Example 8

An electric motor takes a current of 10 A at 240 V and produces a brake power of 2·2 kW. Find its efficiency.

Solution

$$\eta = \frac{P_b}{VI} = \frac{2{\cdot}2 \times 10^3}{240 \times 10}$$

$$\therefore \text{efficiency} = 0{\cdot}917 \text{ or } 91{\cdot}7\%$$

Example 9

A 250 V electric kettle contains 1·75 kg of water at 2°C. It draws a current of 8 A for 7·5 min. in order to boil the water. Find the efficiency.

Solution

$$\begin{aligned}\text{energy output} &= \text{heat transfer to water}\\ &= mc\,\Delta T\\ &= 1{\cdot}75 \times 4{\cdot}2 \times 10^3 \times 98\\ &= 720 \times 10^3 \text{ J}\\ \text{energy input} &= \text{electrical energy supplied}\\ &= VIt\\ &= 250 \times 8 \times 7{\cdot}5 \times 60\\ &= 900 \times 10^3 \text{ J}\\ \therefore \text{efficiency} &= \frac{720 \times 10^3}{900 \times 10^3}\\ \therefore \text{efficiency} &= 0{\cdot}8 \text{ or } 80\%\end{aligned}$$

Example 10

A small water turbine uses 102 kg/s from a reservoir whose level is 20 m above the turbine. A brake test on the turbine, using a rope brake, gave the following data: radius of flywheel, $R = 0{\cdot}6$ m; radius of rope, $r = 15$ mm; spring balance reading, $S = 52$ N; dead load, $F = 228$ N; speed of flywheel, $n = 1500$ rev/min. Find:

i the power input;
ii the brake power;
iii the efficiency.

Solution

i Power input

$$= \dot{m}gh = 102 \times 9{\cdot}81 \times 20 = 20 \times 10^3 \text{ W}$$

ii Braking torque, T_b

$$\begin{aligned}T_b &= (F - S)(R + r)\\ &= (228 - 52)(0{\cdot}6 + 0{\cdot}015)\\ &= 176 \times 0{\cdot}615\\ &= 108 \text{ Nm}\\ \therefore \text{brake power} &= 2\pi n\, T_b\\ &= 2\pi \times \frac{1500}{60} \times 108\\ &= 17 \times 10^3 \text{ W}\end{aligned}$$

iii Efficiency $= \dfrac{\text{brake power}}{\text{input power}}$

$= \dfrac{17 \times 10^3}{20 \times 10^3}$

$= 0{\cdot}85$ or 85%

Exercises

1 2 kg of oil having a specific heat capacity of 2·4 kJ/kg K are heated from a temperature of 60°C to 92°C. Find the heat energy transferred to the oil.

2 1·4 MJ of thermal energy are supplied to a mass of water raising its temperature by 70 degrees. If the specific heat capacity of water is 4·2 kJ/kg K find the mass of the water.

3 It takes 640 kJ of thermal energy to raise the temperature of 50 kg of firebrick from 15°C to 30°C. Find the specific heat capacity of firebrick.

4 4 kg of water at 8°C are mixed with 6 kg of water at 30°C. Find the final temperature.

5 A prony brake had a load-arm 0·4 m long. The load on the arm was 16 N when testing a small engine running at 1500 rev/min. Find the brake power of the engine.

6 An engine produces a brake power of 1·5 kW when tested with a prony brake. The engine runs at 750 rev/min and the prony brake has a load-arm 0·5 m long. Find the load necessary at the end of the load-arm of the brake.

7 During a test on a turbine using a rope brake, the following data was noted: dead load = 200 N; spring balance reading = 40 N; diameter of flywheel = 1·6 m; diameter of rope = 30 mm; shaft speed = 1800 rev/min. Find the brake power of the turbine under these conditions.

8 An engine is expected to produce a brake power of 15 kW when tested with a rope brake. The diameter of the flywheel is 900 mm, the rope has a diameter of 20 mm and the dead load is 180 N. What would be the reading on the spring balance when the engine is running at 2000 rev/min?

9 A diesel engine running at 3000 rev/min produced a brake torque of 128 Nm. The fuel consumed in 20 minutes was 4 kg. Take the specific energy of the fuel to be 36 MJ/kg. Find: **i** the brake power, **ii** the fuel power input, and **iii** the brake thermal efficiency.

10 10 litres of fuel of specific energy 45 MJ/litre are consumed by an engine during a 50 minute test. The engine was found to have a brake thermal efficiency of 26%. Find the brake power.

11 An engine which produces 15 kW as brake power has a brake thermal efficiency of 0·35. It uses fuel whose specific energy is 33×10^6 J/kg. What mass of fuel would be used in a 2 hour period?

12 An engine uses 0·006 litres of fuel per second when developing a brake power of 61·5 kW. If the engine has a brake thermal efficiency of 25% what is the specific energy of the fuel?

13 An electric motor which is 78% efficient develops a brake power of 650 W when supplied from a 240 V source. Find the current taken by the motor.

14 A current of 6 A is supplied to a 220 V d.c. motor which runs at 1380 rev/min. When using a prony brake a load of 20 N was required at an effective radius of 400 mm. Find the efficiency of the motor.

15 An electric immersion heater is 60% efficient. It is supplied with 10 A at 240 V for 45 minutes. During this time it raises the temperature of the water in the tank from 10°C to 74°C. The specific heat capacity of water is $4{\cdot}2 \times 10^3$ J/kg K. Find the mass of water in the tank.

16 A 2·4 kW electric kettle holds 1·9 kg of water at 4°C. If the kettle is 88% efficient, how long will it take to boil the water? (Take c = 4·2 kJ/kg K.)

17 A water turbine is 45% efficient when taking 4000 kg of water per second from a reservoir whose level is 30 m above the turbine. Find the brake power.

18 A rope brake test was performed on a water turbine and the following data was noted: speed of turbine shaft = 2000 rev/min; head of water = 40 m; flow rate of water = 200 kg/s; dead load = 185 N; spring balance reading = 25 N; diameter of brake wheel = 2·4 m; diameter of rope = 20 mm. Find: **i** the braking torque, **ii** the brake power, **iii** the input power of the water, and **iv** the efficiency of the turbine.

Answers

Chapter 3

1 30 km
2 25 s
3 i 6·25 m/s **ii** 375 m **iii** 0·125 m/s²
4 i 30 s **ii** 0·267 m/s²
5 i 5 s **ii** 4 m/s²
6 i 56·3 m **ii** 156 m
7 i 3·5 s **ii** 34·3 m/s
8 12·5 m/s
9 i 22 km **ii** 10 km due East **iii** 189 km/h
10 i 25 m **ii** 20 m **iii** 45 m
iv 6·43 m/s **v** 2 m/s²
11 i 15 s **ii** 10 s **iii** 375 m
iv 15 m/s
12 i 2·5 m/s² **ii** 1·25 m/s² **iii** 0·625 m/s²
iv 210 m
13 i 108 km/h **ii** 1·8 km **iii** 18 km
iv 2·7 km **v** 22·5 km **vi** 90 km/h
vii 0·167 m/s²
14 B wins the race by 20 metres
15 24 N
16 29·4 N
17 1·5 m/s²
18 30 kg
19 3·75 kN
20 62·5 kN

Chapter 4

2a 5 N ↗ 36·9°
b 10·4 N vertically down
c 8·5 N 41·7° ↘
d 5·97 N 76·1° ↘

3a H = 8·66 N → V = 5 N ↑
b H = 3 kN → V = 5·2 kN ↑
c H = 153 N ← V = 129 N ↑
d H = 8·55 kN ← V = 23·5 kN ↓
4a E = 76 N 31·8° ↙
b E = 3·46 kN ↖ 60°
c E = 283 N 45° ↙
d E = 4·74 N ↖ 38·1°

5a	AD = 8·66 kN	Strut	R_x = 7·5 kN ↑	
	BD = 5 kN	Strut	R_y = 2·5 kN ↑	
	CD = 4·33 kN	Tie		
b	AD = 10 MN	Tie	R_x = 21.8 MN	↗ 23·4°
	BD = 17·3 MN	Strut	R_y = 8·66 MN ↑	
	CD = 15 MN	Tie		
c	AD = 981 N	Strut	R_x = 1600 N	37·9° ↘
	BD = 981 N	Tie	R_y = 1260 N →	
	BC = 1260 N	Tie		
	CD = 640 N	Strut		
d	AC = 27·7 kN	Tie	R_x = 46·9 kN	↗ 81·2°
	AD = 27·7 kN	Tie	R_y = 27·7 kN	75° ↙
	BD = 37·9 kN	Strut		
	CD = 27·7 kN	Strut		
e	AD = 20 kN	Strut	R_x = 17·3 kN →	
	BD = 17·3 kN	Tie	R_y = 20 kN	30° ↘
	CD = 0,	Redundant		
f	AD = 2 kN	Tie	R_x = 4·47 kN	26·6° ↘
	BC = 4 kN	Strut	R_y = 4 kN →	
	BD = 2·83 kN	Strut		
	CD = 2·83 kN	Tie		

6a F = 5·5 N
d F = 200 N
c x = 50 mm
d F = 35 N

7a R_x = 70 N ↑ R_y = 80 N ↑
b R_x = 400 N ↑ R_y = 800 N ↑
c R_x = 72·1 N ↗ 16·1° R_y = 60 N ↑
d R_x = 8·53 kN ↗ 70·3° R_y = 4·76 N ↑
8 i 0·8 Nm
ii 64 N

9a $\bar{x} = 27{\cdot}7$ mm $\bar{y} = 33{\cdot}1$ mm
b $\bar{x} = 40$ mm $\bar{y} = 25$ mm
c $\bar{x} = 30{\cdot}1$ mm along OX from O
d $\bar{x} = 29{\cdot}5$ mm along OX from O
10a $\bar{x} = 72{\cdot}7$ mm along OX from O
b $\bar{x} = 47{\cdot}4$ mm along the connecting bar from the smaller sphere (i.e. 12·6 mm from the larger sphere).
11 i Centre of gravity lies on the vertical line which is 5·75 m from O
ii $Fg = 487$ kN per metre length
12 48·8°

Chapter 5

1a 35° **b** 4° **c** 13·5° **d** 5·7° **e** 17·5°
2a 0·81 **b** 0·51 **c** 0·11 **d** 0·4 **e** 0·9
3a 72·8 N **b** 53·6 N
4a 687 N **b** 491 N
5 100 kg
6 $\mu = 0{\cdot}268$, $\phi = 15°$
7a The coefficient of friction for shoes on stone must be the larger, allowing a better grip.
b Shoes on stone floor: $\mu = 0{\cdot}81$, $\phi = 39°$
shoes on wooden floor: $\mu = 0{\cdot}51$, $\phi = 27°$
8a 30·8 N **b** 72·8 N
9a 14° **b** 98·8 N
10 30·3 N at 18° up from the horizontal.
11a 1·16 kN **b** 230 N
12 $\phi = 23{\cdot}3°$, $\mu = 0{\cdot}43$
13a $\theta = \phi = 13{\cdot}5°$ **b** 23·3 N
14 112 N up the plane.
15 4 m ramp, 19·2 N down the ramp.
16 250 N (just to balance the T.R.)
17 5·74° or approximately 1 in 10
18 402 N
19 1·96 kN
20 3 kN
21a 24 N **b** 7·92 Nm
22 80 mm
23 9×10^{-3} Nm
24 2·5 kN

Chapter 6

1 700 J
2 500 kg
3 1·5 km
4 50 N
5 93 kJ
6 44·8 kJ
7 124 kJ
8i 1·5 N/mm **ii** 1·9 J
9i 240 N **ii** 2·4 J
10 2250 J
11i 212 kJ **ii** 115 kJ
12 976 kJ
13 58·9 MJ
14i $2{\cdot}31 \times 10^9$ J **ii** 20
15 120 N
16 75 J
17i 486 kJ **ii** 395 kJ (91 kJ of Work done against friction)
18 0·34
19 11·1 kJ per minute
20 100 N

Chapter 7

1 20 J
2i 25 J **ii** 3·2 N/mm
3 294 kJ
4 15 kJ
5i 8 kJ **ii** 2 kJ **iii** 0·8 or 80%
6 89·6 MJ
7 130 m
8 16 kJ
9 200 W
10 400 W
11 750 W
12i 2·25 kW **ii** 27·3 kW
13 100 N
14 2·2 kW
15i 9·5 kW **ii** 856 kg **iii** 1·13 m/s

Chapter 8

1 MA = 20
2 VR = 5·5
3 Efficiency = 65%
4 Work input = 889 J
5 i MA = 26·7 **ii** VR = 33·3 **iii** W_I = 75 J
iv 60 J **v** η = 0·8
6 i MA = 16 **ii** S_I = 107 m **iii** VR = 35·7
7 i VR = 2 **ii** Efficiency = 98·1%
8 Efficiency = 87·5%
9 Effort = 18·8 N
10 Radius of wheel = 167 mm
11 i S_I = 9·54 m **ii** Efficiency = 65·4%
12 i 18 N **ii** Efficiency = 58·1%
13 i 2·5 N **ii** 9 N **iii** 60%
14 Efficiency = 70%
15 i 2200 W **ii** 1870 W **iii** 89·3 Nm
16 i 225 rev/min **ii** 27·3 N **iii** 28 km/h
17 48 teeth
18 i C **ii** 0·8, 1·6, 2 **iii** 700 rev/min
19 i 7·5 **ii** 100 rev/min
20 30
21 i 0·565 m/s **ii** 848 W **iii** 636 W
iv 1130 N
22 i 22·3 N **ii** 0·118 m/s
23 i 31·4 **ii** 80
24 i 314 **ii** 56·9 N
25 i 0·08 m/s **ii** 80 W **iii** 302 W
iv 26·5% **v** 62·5 s

Chapter 9

1 120 C
2 2 A
3 200 s
4 i 75 Ω **ii** 20 mA **iii** 3V
5 12 V
6 20 mA
7 900 Ω
8 36 V
9 2 mA
10 200 Ω
11 i 110 Ω **ii** 1·8 MΩ **iii** 4·3 kΩ
iv 680 Ω **v** 82 kΩ
12 60 Ω
13 3·3 kΩ
14 680 Ω
15 i 20 Ω **ii** 7·6 Ω
16 2·4 Ω
17 405 Ω
18 2·73 Ω 11 Ω 13·8 Ω 18·3 Ω 30 Ω 7·5 Ω 6·66 Ω 4·1 Ω
19 i 2·97 Ω **ii** 1·6 A
20 i 29·3 Ω **ii** 290 mA **iii** 53 mA 85 mA 152 mA
21 i 2 kW h **ii** 7 p
22 i 0·2 kW h **ii** 0·16 kW h **iii** 0·04 kW h
iv 75 times
23 12·5 A
24 i 5 kW **ii** 7·5 kW
25 i 120 W **ii** 1·2 Ω
26 i 5 A 2·4 Ω **ii** 2 A 6 Ω **iii** 0·5 A 24 Ω
iv 0·25 A 48 Ω
27 i 0·625 A **ii** 384 Ω **iii** 0·9 kW h
iv 4·5 p
28 i 500 Ω **ii** 0·48 A **iii** 73·9 V
iv 0·127 A **v** 0·274 A
29 i 40 Ω **ii** 3 A **iii** 0·54 kW h
30 A_1 = 2 A A_2 = 1·25 A A_3 = 1·5 A

Chapter 11

1 $x = 0{\cdot}6 \times 10^{-3}$ mm
2 F = 600 N
3 σ_A = 113 MPa and σ_B = 50 MPa
4 A = 20 mm^2
5 F = 33·9 N
6 $\epsilon = 0{\cdot}15 \times 10^{-3}$
7 x = 5 mm
8 l = 800 mm or 0·8 m
9i σ = 69 MPa **ii** ϵ = 0·0006 **iii** E = 115 GPa

10 $\sigma = 29{\cdot}7$ MPa
11 $\epsilon = 2{\cdot}85 \times 10^{-5}$
12 $A = 1{\cdot}96$ mm^2
13 $F = 216$ kN
14 f.o.s. = 7
15 $A = 24$ mm^2 (minimum)

Chapter 12

1 $Q = 154 \times 10^3$ J
2 $m = 4{\cdot}76$ kg
3 $c = 853$ J/kg K
4 $t = 21{\cdot}2$°C
5 $P_b = 1{\cdot}01$ kW
6 $F = 38{\cdot}2$ N
7 $P_b = 24{\cdot}6$ kW
8 $S = 24{\cdot}3$ N
9i $P_b = 40{\cdot}2$ kW **ii** $P_i = 120$ kW **iii** $\eta_{bth} = 33{\cdot}5\%$
10 $P_b = 39$ kW
11 $m = 9{\cdot}35$ kg
12 Specific energy = 41 MJ/litre
13 I = 3·47 A
14 $\eta = 87{\cdot}6\%$
15 m = 14·5 kg
16 time = 363 s or 6 min 3 s
17 $P_b = 530$ kW
18i $T_b = 194$ N m **ii** $P_b = 40{\cdot}6$ kW **iii** $P_i = 78{\cdot}5$ kW
iv $\eta = 0{\cdot}517$

Index

Index